To

Mr. & Mrs. John Solo

from

Bob Thomas

Christmas 1980

P9-CFA-531

First Person Rural

FIRST
PERSON
RURAL

ESSAYS OF A SOMETIME FARMER

BY NOEL PERRIN

DAVID R. GODINE ◦ PUBLISHER ◦ BOSTON

First published in 1978 by
David R. Godine, Publisher
306 Dartmouth Street   Boston   Massachusetts

ISBN 0-87923-232-3   LCC NO. 77-94109

ACKNOWLEDGMENTS: 'Barter,' 'The Grades of Maple Syrup,' 'Real Milk,'
'Grooming Bill Hill,' 'Making Butter in the Kitchen,' 'A Cool Morning in
Vermont,' 'Market Research in the General Store,' 'The Two Faces of Ver-
mont,' and 'Flow Gently, Sweet Maple' are reprinted by permission of
Vermont Life. 'Sugaring on $15 a Year,' 'Buying a Pickup Truck,' 'Buying a
Chainsaw,' 'In Search of the Perfect Fence Post,' and 'Raising Sheep' are reprinted by permission of the Country Journal. 'The Wooden Bucket Principle,'
'Tell Me, Pretty Billboard,' and 'Old MacBerlitz Had a Farm' are reprinted by
permission of The New Yorker. 'Selling Firewood in New York' is reprinted
by permission of New York. 'The Other Side' is reprinted by permission of
the New York Times. Five of the titles have been changed.

FIRST PRINTING June 1978
SECOND PRINTING August 1978
THIRD PRINTING February 1979

Printed in the United States of America

FOR ANNEMARIE

Foreword

IMAGINE A LARGE FAMILY GATHERING, something like the one Robert Frost describes in 'The Generations of Men.' Here are a couple of cousins who have never met in their lives before. One is a teacher in New York (with a summer place in the Berkshires); the other has lived in Vermont all her life. Over yonder are four bearded brothers, talking as busily as if they hadn't just seen each other at the Grange meeting three days ago. The fellow in the plaid pants is Uncle Philip, from Hartford, Connecticut. He sells life insurance–belongs to the Million Dollar Club. He is talking to Cousin Sarah, who raises hens.

This book is a little like such a gathering. The essays in it come from many places, and range from intensely practical to mildly literary. One large group is a collection of columns from *Vermont Life*, where they appeared under the title 'The Part-Time Farmer.' Several others come from Blair and Ketchum's *Country Journal*; and all of them are about part-time farming, too. On the other hand, two or three pieces are from *The New Yorker*. None of them would be helpful to a farmer. One pretends to be about wooden sap buckets, but is really about reality and illusion. Another reports on the sounds said to be made by cows, pigs, and various other animals in different parts of the world. The only thing all the essays have in common–what makes them a family– is that they are all concerned with country-ish things.

I have revised several of the how-to-be-a-farmer pieces in the light of experience I've had since writing them, and I've added a long postscript to one rather disenchanted article on the bad side of country life. I've added a shorter postscript to one of several pieces about maple syrup. The rest have merely had their faces washed and their hair combed, prior to meeting their cousins in this book.

NOEL PERRIN
January, 1978

Contents

First Person Rural

Barter

A MODERN FARMER is usually as much into the money economy as any bank president. Even in Vermont, the real farmers are constantly converting farm products into dollars. The big event of the month is the arrival of the milk check. If a farmer sells off some timber, it's a cash deal with New England Stumpage, Inc. With some help from the county agent, he computerizes his farm accounting.

But amateur farmers like myself can still enjoy the pleasures of trading. Most of the time I'm as cash-minded as the next man. But farming, I hardly soil my hands with money from one month to the next. This past year, for example, my unit of trade has been the fence post.

It all began on Christmas Day, 1974. While my wife was in the final three-hour plunge of fixing dinner, I went out for a walk. (Before anyone leaps to conclusions, let me add that she had sent me out to walk. She likes her kitchen undisturbed before major feasts.)

My walks usually have a purpose. This time I was looking for three or four long poles to use in a new barway I was planning. Snowshoeing up the back hill, into a little worthless corner of my land, I found a stand of hemlock that in ten years I had somehow never noticed before. There were a dozen big ones, a couple of feet in diameter, and then several hundred smaller ones, crowded so close together that they were killing each other. They ran

about 30 feet tall, and were four to eight inches in diameter at the butt. In the middle was a 200-year-old maple they were busy shading out.

I cut four good poles in no time. Then I began looking at that threatened maple, and at the half-dead hemlocks scattered through the grove—and without ever consciously making a decision, I found I had taken on my next project. I was going to thin that stand.

Only, what was I going to do with the trees I was about to cut? I didn't need any more poles. Hemlock is next to worthless as firewood. To cut a quarter of these young hemlock and leave them to rot might make business sense, but it would offend the instinct that led me to part-time farming in the first place.

Then I realized that I was looking at a couple of hundred potential fence posts. I had absolutely no plans to fence anything for the next year or so. But so what? I'd make my posts and then see what happened. If nothing, well, I have a big barn and plenty of room to store things.

By the end of the winter, I had an enormous pile of six-foot posts, all sharpened at the larger ends, plus a much improved stand of hemlocks. I had also already made my first trade. I had dropped off thirteen posts at the house of a neighbor who was planning to re-fence his chicken run. In return, I brought back four bantam hens, at three posts each, and a rooster for one. (His place is crawling with bantam roosters, and he probably would have let one go for free. But it's no fun to trade that way.)

All last spring, the posts just sat around, though one day I did lug 30 or 40 in and stack them in the barn. Then, in June, I started a new project, which was to enlarge the woodshed. I needed some eight-foot pine boards for the new piece of roof. I *could* have gone to a lumberyard and bought some. But I didn't. Instead I called an old friend about 30 miles away who had just bought a little sawmill. We quickly made a deal whereby the next time I came down to visit I'd bring along a few eight-foot pine logs, plus twenty fence posts for him. He would saw me out 200 feet of boards in return.

As it turned out, I miscalculated. My four logs, though they cost me half a day to cut and to load in the pickup, only yielded about 160 feet of boards. Sixteen posts' worth. But fortunately Willis's cow had just freshened, and we completed the deal with two gallons of milk, at two posts the gallon.

This past winter I made the best deal of all. I am a second-class or perhaps even third-class chainsaw sharpener. I can touch up a dull chain handily enough, but I can no more adjust the bite of the teeth and straighten the angles than I could handle the orthodontistry for my daughters. It has been a source of embarrassment and humiliation to me ever since I first bought a chainsaw that I have to take it back to the dealer whenever it needs a major filing. Once I bought a completely superfluous new chain just to put off the evil day. But now I have an arrangement with another part-time farmer, a couple of miles away. I can get the chain filed as good as new whenever I want, at two posts a time. I have been walking straighter and looking people more directly in the eye ever since.

The Grades of Maple Syrup

MAPLE SYRUP comes in three grades. In New York State they are called Light Amber, Medium Amber, and Dark Amber. In Vermont they are called Fancy, A, and B. I am speaking, of course, of pure maple syrup: The well-known 'blends' that are sold in supermarkets (at the moment most of them contain 3% maple syrup) come in one grade only, called Mediocre.

Of the two systems, Vermont's is clearly better. When you know that the syrup you're about to buy has been classified Medium Amber, you don't know much. You might as well grade meat Pink, Dark Pink, and Red. But even Vermont's system no longer serves the consumer well, though it once did. To understand why, you need to know a little of the history of maple marketing.

A hundred years ago, most Vermont farms produced only maple sugar, and no syrup whatsoever. Syrup would have been too hard to ship. Tin cans had been only quite recently perfected by the brothers Appert in France and were expensive. Plastic jugs didn't exist. The common shipping container in Vermont was a barrel, keg, or box made of wood—and maple sugar left the farm chiefly in wooden boxes. I know a man in East Corinth whose grandfather made his living producing boxes for farmers to ship their butter and maple sugar off to market in. He produces boxes still—but now they are hat boxes and miniature trunks to give little girls to pack doll clothes in.

Furthermore, a hundred years ago maple sugar was not a luxury item. It was competitive in price with cane sugar. Cane sugar in the 1870's sold for about 7¢ a pound, and maple sugar sold for an average of 9¢ a pound–a farmer in Cabot sold four tons of it for that price in 1878. Occasionally maple sugar even undersold cane; another Cabot farmer sold all of his last-run sugar at 6¢ a pound that year.

Maple sugar was also used competitively with cane. That is, the buyer expected to use it as a general sweetening agent–in his coffee, with strawberries and cream, in cake recipes, and so on. For this reason it was sold almost entirely in bulk, just as cane sugar is now. A young sugarmaker in northern Vermont was saying something revolutionary when he wrote in 1886, 'I have learned that small packages generally sell the best, those containing from ten to 30 pounds finding the quickest market.' Think what the large packages must have been like. By contrast, a small package of maple sugar in 1978 is likely to be four ounces, and a large one half a pound.

So much was maple sugar considered a general sweetening agent that a certain number of Vermont chauvinists (they have existed in all ages) felt that none of that Cuban or Louisiana cane stuff need be tolerated at all. A fellow in Bakersfield said flatly in 1876, 'There ought not to be a pound of foreign sugar brought into the State.'

Now we come to the point. If maple sugar is competitive with cane, and if it's used as a general sweetener, the last thing you want is for it to have a lot of maple flavor. You just want it to be sweet. Maple-flavored coffee *may* be good, but it's not what most coffee drinkers are after.

And, in fact, this is what Vermont producers (and also New York producers, and the early Wisconsin producers, etc.) were trying to make: colorless and hence flavorless maple sugar. That is, the very fanciest Fancy Grade.

'There is no good reason why we cannot make the Maple equally white and pure as the West Indies,' one of the big producers in South Reading said in 1878. A somewhat more realistic farmer in Waitsfield didn't think he could boil his sap down to a sugar without *any* flavor or tint—but he thought it was a goal one should approach as closely as possible. Speaking of his own maple sugar, he wrote, 'Like the human race it is of all shades of color, and I think this is one of the cases in which prejudice against color is justifiable. We have all seen maple sugar that was nearly as white as loaf sugar, and I suppose all would be glad to make it.' All would be glad to make it because the housewife in Boston or New York would pay the highest price for it.

Here is the origin of the Fancy, A, B grading system, and also of the Light Amber, Medium Amber, Dark Amber. Fancy means palest and mildest flavored—what you make at the beginning of each season. A means somewhat darker and more flavor. B means still darker and still more flavor. (Ungraded or c is, of course, darker yet—and really dark c is usually too strong to use by itself with any pleasure. That's why so much of it winds up in supermarket blends.)

This system made perfect sense in 1878. In 1978, when virtually all maple syrup and sugar is sold *because* of its flavor, it makes very little sense. I won't say it makes no sense at all, because the differences between the three grades are not simply matters of intensity. To most palates, the pale Fancy grade has a subtlety and delicacy that B completely lacks. To most palates, B has a kind of full-bodied quality, a robustness, sort of like a Burgundy wine, that Fancy completely lacks. A is a brilliant compromise, subtle but sturdy.

Nevertheless, to use a scale on which palest is always best seems to me silly. Here are three kinds of maple syrup, each a good thing in its own right. I myself prefer Fancy on plain raised doughnuts at a sugar-on-snow supper, A on vanilla ice cream and on waffles, B on pancakes—though I also like to switch them around occasionally. But the casual buyer, seeing all three kinds together, figures that A must be an inferior version of Fancy, and B an inferior version of A. Whereas if you kept the same divisions, but called the three grades Mild, Medium, and Strong, the casual buyer would know what he was getting.

9

On the other hand, maybe after a hundred years the tradition is too engrained, and the present grading system cannot be changed. After all, we still call ourselves sugarmakers, and groves of maples sugar orchards, and the buildings that house our evaporators sugarhouses, when for three-quarters of a century we have been syrupmakers tapping our syrupbushes and boiling down the sap in our syruphouses. In that case, it seems to me that maple syrup cans should at least carry an explanation of what Fancy, A, and B mean.

Too many people are missing a treat, not putting some wonderful dark rich B on their pancakes from time to time.

POSTSCRIPT. Since I wrote this, I have learned that maple grading is even more complicated than I had realized. Vermont is still grading Fancy, A, and B; and New York is still using Light Amber, Medium Amber, and Dark Amber. But it turns out that Canada has *five* grades. All syrup that Vermont grades at all is in Canada called Class 1–and divided into three sub-classes called Extra Light, Light, and Medium. These three exactly correspond to the Vermont and New York grades.

But the end-of-the-season runs that in Vermont are unofficially called Grade C get divided into two official classes in Canada. The better-quality C (which is in fact good stuff, though powerful) they call Class 2, or Amber. The buddy-tasting stuff that you get at the very end of the season they call Class 3, or Dark. That, I suspect, is where the 3% in Log Cabin, etc., comes from.

But things are even more complicated than that. Recently the U.S. Department of Agriculture has gotten into the act. The Department wants to create an all-American grading system. Their idea is to call everything Grade A except Grade C. That they would call Grade B. The new inclusive Grade A would then be divided into three categories. Guess what they would be called. Light Amber A, Medium Amber A, and Dark Amber A. I suspect the whole thing of being a plot hatched by the giant food conglomerates and designed to confuse not only people who buy real maple syrup but the producers as well.

Real Milk

A FEW PEOPLE keep a cow. Most people go to the super-market. The typical housewife comes back with half a gallon of pasteurized, homogenized milk (for the kids), a quart of pasteurized, homogenized low-fat milk (for herself), and a pint of thin "cream" called something like Coffee-Cereal Special (for the whole family). Maybe a pound of butter. About twice a year she gets half a pint of heavy cream for whipping. No oftener, because it's too expensive.

Americans tend to think these are the only choices. Either you do your own milking twice a day, or else you buy highly processed dairy products from the rack.

In most of Vermont, there's a third choice. You can buy whole milk direct from a farmer–and make your own butter, your own low-fat milk, your own virtually free whipped cream, and so on. As fringe benefits, you decrease pollution, and maybe save some money. Besides, it's fun to make butter.

My wife and I have been getting our milk this way for the last two years. We buy it from a neighbor who used to work on the state highway crew. Now that Floyd is retired, he's gone back to farming. His real business is beef cattle, but he keeps one milk cow, currently yielding eighteen quarts a day. He has enough to supply his own family, three customers, and his pig. (The pig could use more–but that's what it means to be a pig.)

It works like this. We have a supply of quart glass bottles, which originally held Tropicana orange juice. Floyd has an old refrigerator down cellar. Every two days we go by and pick up four quarts of milk from that refrigerator, at the same time returning washed bottles from last time. I have read that commercial returnables are good for about ten round trips—but home-use ones do better. Ours have lasted several hundred trips. Only one broken bottle so far. We pay $1.25 a gallon.

Obviously the milk we buy is not pasteurized. It's so-called raw milk. I'll explain later why I think that name is wrong. Which does raise some questions.

First, is it really legal for Floyd to sell unpasteurized milk to us? Yes, it really is. Under Vermont law, any farmer can sell up to 25 quarts a day without a license. With a license—which means state inspections and fees—he can sell as much as he pleases. There are two dairy farms in the state that sell their entire production that way. Peter Smith at the Old Nash Farm in Middlebury sells about 60 gallons of unpasteurized and unhomogenized Jersey milk a day. Eric Help in Montpelier, who just started a year and a half ago, is selling about 35 gallons a day.

The one rule is that anyone who sells milk in Vermont must charge the minimum price set by the Vermont Milk Control Board. At the moment I write, it's $1.48 a gallon. This is exactly what Peter Smith does charge. Eric Help's milk, also Jersey, costs $1.60 in Montpelier, and quite a lot more in the health-food restaurants that stock it. But nearly all the farmers who sell just a few gallons to their neighbors, like Floyd, charge either $1.00 or $1.25. Beyond pointing out to them how foolish they are, the Milk Board doesn't do much about it.

Second question, is it really safe? Yes. Milk was originally pasteurized for two reasons: because of the danger of tuberculosis, and because of undulant fever. TB from cows has all but vanished. Dr. Philip Nice of the Dartmouth Medical School says he has never yet seen a case of bovine tuberculosis in this region. (But Vermont cows are all tuberculin-tested, anyway.) As for undulant fever, or brucellosis, that does occasionally crop up. There was at least one case in Vermont last year. A dairy farmer. But he got it from actually handling cattle with brucellosis. He and his whole family were drinking unpasteurized milk from those same cows—and he was the only one to get sick. The risk from drinking milk is pretty low.

If you ask why take any risk at all, I answer that it is impossible not to take any risk at all. Life isn't like that. You take a risk drinking supermarket milk, too. Remember, it's homogenized. Some doctors think that's bad for your health. What homogenizing does is to break up the natural milk fats into microscopic globules—which they think increases the risk of heart disease. And what pasteurizing *doesn't* do is get rid of all bacteria. It gets a lot, but there are always survivors. In fact, Peter Smith's unpasteurized milk regularly has a lower bacteria count than your average carton of supermarket milk.

Maybe give up milk altogether? You risk calcium problems. Synthetic milk? There's mounting evidence that synthetic dairy products, which are mostly based on soybeans and cocoanut oil, have risks of their own.

Actually, if you wanted to reduce your risks to the absolute minimum, you'd get your own cow. Or else buy direct from a farmer. That way you'd avoid homogenizing. Then you'd buy a home pasteurizer—they're not very expensive—and thus avoid the minute risk of undulant fever. This is what our new family doctor and his young farmer-wife do. We don't, because we prefer the flavor of fresh milk.

Which brings up the last question: What does unpasteurized milk taste like? Well, like pasteurized milk, only more so. There is no question that it has more flavor, just as fresh tomatoes have more flavor than stewed ones. Or fresh lettuce than boiled lettuce. These are fair comparisons, which is why I don't think "raw" milk is the right name. Raw is what uncooked meat is—with the strong implication that it *ought* to be cooked. And so it should. But milk is a finished product as it comes from the cow. One might as well speak of raw orange juice, or a breast-fed baby guzzling raw mother's milk.

I don't claim that more flavor is always a good idea. As a matter of fact, all three of our children refused to drink farm milk at first, because they said they could taste it. One of them still won't touch it. On the other hand, most guests who try it like it a lot; one has since bought a Guernsey heifer.

Grooming Bill Hill

QUESTION: Why is Vermont more beautiful than New Hampshire? ANSWER: Because of Vermont farmers. Remove the farmers, and within ten years New Hampshire would surge ahead.

This is a serious argument. If you just consider natural endowment, the two states are both fortunate, but New Hampshire is more fortunate. It has taller mountains, it has a seacoast, it even owns the whole northern reach of the Connecticut River, except a little strip of mud on the Vermont side.

But New Hampshire's farmers mostly quit one to two generations ago and started running motels or selling real estate. The result is that most of New Hampshire is now scrub woods without views. Dotted, of course, with motels and real estate offices.

A lot of Vermont farmers, however, are holding on. Almost every farmer has cows, and almost every cow works night and day keeping the state beautiful. Valleys stay open and green, to contrast with the wooded hills behind them. Stone walls stay visible, because the cows eat right up to them. Hill pastures still have views, because the cows are up there meditatively chewing the brush, where no man with a tractor would dare to mow. (That's the other argument for butter besides its taste. I once figured that every pound of butter or gallon of milk someone buys means that another ten square yards of pasture is safe for another year.)

Until lately, my own contribution to the beauty of Vermont was modest. I did fence two little hayfields a few years ago, so that my neighbor Floyd Dexter could run beef cattle there after the hay was cut. Sometimes I run a couple myself. But both of these were good fields when I bought the place. My contribution was merely turning them from straight hayfield to hayfield-that-gets-grazed, so they would stop shrinking a little every year, and so that the cows would eat right up to the stone walls.

This year, however, I think I have seriously joined the ranks of those who maintain Vermont. Or maybe not so much joined as been quietly drafted by Floyd.

It all began because of Bill Hill. Bill Hill is a large lump of glacial debris behind the pasture across the road. I own it. Insofar as a thing as small as a human being can claim to own a thing as big as a hill.

Sixty years ago, it was all pasture. No trees except for one white birch on top, and a row of immense old maples on the slopes behind it. But just before World War II a New York lawyer bought this farm. He naturally kept no cows on Bill Hill. When I got it, one end was completely grown up to woods, and the rest was in every possible stage from briar-choked pasture to almost-woods. The top remained open, and because I like to picnic in a place with a 360-degree view, I have painfully kept it open by dragging a little sicklebar mowing machine up there every couple of years.

Last summer, though, I was watching Floyd's cattle uncover yet another stone wall in the field behind the house and trim the apple trees up perfectly to a height of five feet, and it struck me that there was a better way to maintain Bill Hill than dragging little machines up it. At that time my idea was just to fence four or five acres: the face of the hill we see best from the house, and the top.

The next day Floyd was over looking at a newborn calf, and I told him my idea. He liked it. Together we climbed Bill Hill and tentatively set the bounds. It turned out to be more like seven acres than five, because he pointed out that by using just a little more wire, I could include quite a lot more of the hill.

All winter when I had a spare afternoon I would go over and prune up bull pines and cut out poplars in the pasture-to-be, so as to encourage the grass. I got quite skillful at skiing out with a chain saw in one hand. Floyd got us a couple of hundred cedar posts at East Thetford Auction to supplement my remaining hemlock, and we bought wire at a remarkable store in Topsham called Freddy Miller's.

This spring, as soon as the frost was out of the ground, we began to drive posts. Also to enlarge the boundaries. The very first day we were out, Floyd led us as if by accident through a beautiful level patch of grass just beyond Bill Hill—and before I knew what happened I had agreed to fence nine acres instead of seven.

The boundaries stayed set for about a month. (We were fencing only on weekends, and not all of them.) Then one warm May afternoon, just as we were coming over the hill with the wire, almost ready to turn and close the pasture, Floyd remarked that it was thirsty weather. 'I don't suppose there's any water back here,' he said as we wiped our sweaty faces. I said no, not a drop.

We drove a few more staples in silence, and then Floyd remarked almost dreamily that he had gotten his feet wet deer-hunting behind the hill last fall. Probably dry there now, he added.

'I don't see that,' I said. 'If there was water there in November, there's certainly going to be in May. Let's go look.'

Floyd was skeptical, but just to please me, he came. Sure enough, about 200 yards beyond where I had meant to turn the fence, there was a good-sized wet place right near my boundary wall with Ed Paige, and even a tiny stream running. In thirteen years, I had never noticed it. Too grown up with briars and brush.

'Awful good to have water where you want the cattle to graze,' Floyd said. 'It'll keep them out on the hill. Course, this probably dries up along about June.' As he spoke, he was walking steadily uphill from the wet spot to a place where someone had rocked in a spring, probably 150 years ago. People don't do that for places that dry up in June. We dug it out a little with our hands, let the water clear, and had a drink. I had never seen the spring, either. Floyd knows my land better than I know it myself.

Since the whole idea is to keep the cattle on the hill, I didn't even much resist taking the pasture on back, even though I had now committed myself to fifteen acres. And it was my own idea –Floyd wasn't even present–when I decided the next day to go back still further, to the stone wall by the maples, and turn the wire down that.

That's how I come to be adding eighteen acres of pasture this year. That's how come for the next half-century, at least, there will be one green grassy hill in Thetford Center, Vermont, to contrast with the dozen or so wooded ones, and a new green meadow behind it. There will be cows against the skyline, and there will be four new stone walls visible. It will be no bad legacy to leave.

Making Butter in the Kitchen

ABOUT SIX MONTHS AGO, my wife and I were on our way home with four quarts of unpasteurized, unhomogenized milk. We had gotten them (as we still do) from our neighbor Floyd, who keeps one Guernsey along with his beef cattle.

Leaving our car in the barn, we went through the shed and into the kitchen. While I put some wood in the stove, Annemarie took a ladle and dipped most of the cream off the top of each bottle. She'd really like to take it all, but the children and I resist, so she takes about two-thirds. When she had finished, she had a little over a pint of beautiful rich thick heavy cream, and three and a half quarts (distributed evenly among four bottles) of low-fat but distinctly not skim milk. Now to make butter.

Because we were going out to dinner that night, and because Annemarie's favorite present to take a hostess is a little pot of freshly made butter, she went into production immediately. Because I was writing this account, I stood about two inches behind her, watching and getting in the way.

Up until two years ago, when Annemarie started her home dairy, I had a completely nineteenth-century image of the process. In my mind's eye I saw a grandmotherly-looking woman seated at a kitchen table, turning a large wooden churn. And turning it, and turning it, and turning it. My mind even supplied two or three strips of flypaper hanging down over the table.

What my blonde wife does in our flyless kitchen is less picturesque, but a good deal faster. She begins by pouring most of the cream into a blender, which she sets on high speed. While it's running, she pours the bottom–and hence thinnest–quarter of the cream into a pitcher, and puts it in the refrigerator. That's our coffee cream.

So far the whole process has taken about 90 seconds. The cream in the blender has become a sort of very thick whipped cream and isn't budging. Annemarie snaps the blender off, and sticks a spoon into the whipped cream. A big bubble of air comes out. It is, in fact, rather like burping a baby.

Then she starts the blender again, on low. The whole mass turns for a few seconds, and then stops. She flicks off the blender, puts her spoon in, and releases another air bubble. She may repeat this process eight times or eighteen times. It depends on how cold the cream is, what the weather is like (I'm serious), and no doubt on other variables we're not even aware of. But it has never yet taken more than five minutes from the beginning of the process to the arrival of butter. You know it's arrived because suddenly there's a lot of loose buttermilk surging around in the blender, and on top of it a soft mound of pale yellow something.

The next step is straining. Annemarie reaches in a drawer, pulls out a clean cloth, and lays it over a saucepan. Then she empties the blender through the cloth, gathers up the corners, and twists lightly. Then harder. There is now a lump of bright golden butter in the cloth, shaped like a giant teardrop. To our surprise, the butter stays golden right through the winter, even though Floyd's cow is then living entirely in the barn and eating entirely hay. Parkay isn't the only one that can do miracles. And Floyd's cow doesn't even use carotene dye.

If we were going to keep the butter ourselves and use it at the next several meals, Annemarie would now be practically finished. But since she's going to give it away and wants it to last almost the way store butter does, she has a couple of more steps to take. First she rinses the cloth under cold water, squeezing even harder. Then she opens it, and 'works' the butter with her spoon for about fifteen seconds. A few more drops of buttermilk appear, which she rinses off. Then she shakes a little salt on, and kneads the butter once more. Then she puts it in a pot, and ties a ribbon around it, ready to give our hostess. But not before I have reached in with a knife and removed one good lump to have with French bread. It's not nearly as important to have butter ten minutes after making it as it is to eat corn ten minutes after picking it. But there *is* a perceptible difference.

The total time elapsed has been just over nine minutes. The total product–I weighed it–is just under five ounces of butter. We paid Floyd $1.25 for the gallon of Guernsey. After Anne-marie's processing, we now have three and a half quarts of low-fat milk, a pitcher of coffee cream, one stick of butter and a little more. All tasting wonderful. And no added chemicals, no wasted packaging material of any sort. Match that, National Dairy Products! Match that, Mazola!

Sugaring on $15 a Year

MOST COUNTRY DWELLERS in New England sooner or later think about doing a little maple sugaring. About nine-tenths of them never actually get around to it. They don't have enough trees, or they don't have enough time, or they don't have the $700 that even a small evaporator costs. Retired people with time and maples and $700 generally don't have the stamina you need to keep slogging through the snow with full sap buckets.

If you are such a frustrated maple sugarer, I have a solution to offer whereby you can sugar this spring with no physical effort and for a total investment of roughly $15. In your spare time. Without setting one foot in the snow.

The trick, of course, is that I am using "sugaring" in its old and true sense—not to mean the production of maple syrup in an evaporator, but the production of maple sugar in a pot. This can be done starting with sap, of course, but it can also be done starting with existing syrup, any old syrup, which is the method I am proposing. Assuming you already have a kitchen with a stove in it, you need only three things to start sugaring: half a gallon of low-grade maple syrup ($7 or less); a rubber mold (about $8); and a touch of skill (free). With these simple ingredients you can turn out several pounds of really stunning maple candy.

Sometimes when people make the kind of claim I have just made—that, with no training and with practically no expenditure of time or money, you can do some wonderful thing—they are secretly expecting you to provide the wonderfulness yourself. I once bought a book on building stone walls, lured by a dust jacket which promised that with just the rocks lying around my fields, I could build handsome retaining walls, set stone steps in them, design beautiful stone culverts, and so on. All this was true. I could have–if I had a natural genius for setting stones. I don't. I can lay up serviceable stone walls, and after ten years of practice, that remains the limit of what I can do.

But making maple sugar is different. It really takes the merest touch of skill. You can be the sort who bends nails, even the sort who breaks the yolk without meaning to when you fry an egg, and still make maple candy that elicits actual moans of pleasure from those who eat it. Maybe not with your first batch, but certainly with your third or fourth.

Let me assume you are now convinced and dying to start. First, obviously, you've got to get some syrup and a mold. The mold you can pick up wherever maple sugaring supplies are sold, which means about one hardware store in three in northern New England. Or you can order one by mail from the Leader Evaporator Company in St. Albans, Vermont, or from the G. H. Grimm Company in Rutland. The most fashionable mold at the moment seems to be one that makes large and realistic maple leaves–I have one myself–but what I recommend for starting is a less complicated mold. Leader has one that makes fifteen hearts that I like pretty well, although I wish the hearts were smaller. (No child I know agrees.) It sells for $8.50.

As to the syrup, get the cheapest you can find. There is no *harm* in its being this year's Fancy Grade, but there's no need, either. Let me tell two stories to show just how low you can stoop.

The first involves my own introduction to making maple sugar. I got into it because of a small disaster. In the spring of 1975 I was making maple syrup as usual and managed to burn a batch of nearly two gallons just as I was about to draw it off. Since I only make about 25 gallons a year (I get tired, slogging through the snow with sap buckets), this meant goodbye to a sizable chunk of the year's production. Burnt maple syrup is not usable; you can filter out the thousands of flakes of black carbon, but you can't filter out the taste.

Fortunately, a friend named Tom Pinder came into the sugar-house before I got around to throwing the stuff out—in fact, while I was still jumping up and down and swearing.

Tom can find a solution for almost anything. If he had been present at the burning of Gomorrah, he would either have figured out a way to put the fire out, or at least had the city rebuilt in no time. He took a look at my scorched pan, and then tasted the contents. 'I've heard,' he said thoughtfully, 'that you can get the burned flavor out of syrup by taking it down to sugar.' And he offered to help me try. I had never before even considered making sugar, because I thought it was too hard for someone of my limited skill.

As I've already suggested, it was easy. To my amazement the burnt flavor vanished almost completely, as did the dark color. We made pie tins full of sugar and cookie pans full of sugar, and finally we bought a mold and made maple-sugar leaves; our friends and families gobbled up what we didn't eat ourselves.

By early summer I had used up the whole two gallons, and I began looking around for a fresh supply. I recalled that back in 1974 another friend, Alice Lacey, had made six or seven gallons of dark Grade B late in the season and had never gotten around to canning it. It was still sitting in her mud room in a ten-gallon milk can. It sported a thick layer of green mold on top. It was now fifteen months old.

Alice was happy to trade me a gallon for a few fence posts, and when *that* syrup produced pale, delicious maple candy, I knew I was onto something. I promptly traded for three more gallons, and then Alice acquired a mold and turned the rest into sugar herself. We have both saved end-of-the-season syrup for sugar ever since.

The process we both use works like this. We pour about a pint (no need to measure) of syrup into a good-sized pot so that it's no more than an inch deep. Half an inch is better. By instinct rather than for any scientific reason I know of, I use a stainless steel pot rather than aluminum. Tom, who hears a lot, has heard that you should always use a wooden spoon rather than a metal one; since I have a wooden spoon, I do that, too. But here I'm not even prompted by instinct. I just like tradition.

You bring the syrup to a boil, and you then turn the heat well down, since even shallow syrup will boil over with extraordinary rapidity. In ten minutes or less you should have the right consistency for sugar. You can test with a candy thermometer or with a special maple-sugaring thermometer, if you like. Leader and Grimm both sell them. I find it simpler to use the hard-ball test: lift your wooden spoon and when the drops are coming off individually let one fall into a glass of cold water. If the sugar is ready, the drop will instantly form into a compact flattened ball in the bottom of the glass. If it isn't ready, the drop will spread out in a sort of little nebula.

You now take the pot off the stove and stir while it cools. To hasten the process, set it in cold water for 30 seconds or so, still stirring. With this shortcut, the sugar usually begins to crystallize within two or three minutes. But it is by no means cold; it is still too hot to touch.

The crystallizing is rather dramatic. What you have been stirring is a thick, opalescent brown taffy, incredibly sticky and tooth-pulling. It now begins to lighten in color and to get rapidly thicker still. Don't worry, just keep stirring. It will then abruptly set.

At this point, dash back to the stove (you have been stirring while comfortably seated at the kitchen table), and put the pot over medium heat. The still-hot sugar will re-liquefy within a minute or so, still keeping its crystalline structure. In fact, at this point the crystals usually get finer and the color a still paler and more elegant tan.

Keep the pot on the stove for another half minute after the sugar re-liquefies, stirring steadily. The sugar is now hot enough to stay pourable for several minutes, and you stroll back to the kitchen table and fill your mold. Since a pint of syrup makes al-

most a pound of sugar, you will have a little left in the pot as a
starter for the next batch. So don't scrape or wash the pot; put it
and the spoon away as is until the next time you sugar. Then add
another pint of syrup and start boiling.

Meanwhile, you have fifteen maple-sugar hearts or twelve maple-sugar leaves, or whatever, sitting in your mold. You could
even have letters spelling out the names of your children, since
any supplier can get you an alphabet mold. Leave them there for
ten or fifteen minutes so you won't burn your hands when you
press them out. Then put one out to eat, still warm, and the
others on a plate to cool. Be sure they get consumed within
roughly the next two weeks, because after that the little cakes
gradually dry out and get hard. I don't think you'll find this a
problem. My difficulty has always been to keep everything from
getting eaten the first day.

There are a good many variations I haven't gone into. Koreans
are said to have twenty-some different names for cooked rice,
depending on how much water they cook it with and hence how
hard it is. Vermonters have only four names for different degrees
of hardness in maple candy: maple cream (which is too soft for
a mold), soft tub sugar, hard tub sugar, and cakes. But in actual
fact, there are infinite gradations from soft tub to cake, and each
tastes just a little bit different. All taste wonderful. This is why I
predict that when you've run through your original half gallon
of syrup, you'll be out scouring the countryside, looking for a
gallon of Grade B to buy cheap.

Buying a Pickup Truck

ONE OF THE WAYS a newcomer to the country knows he's getting acclimated is when he begins to notice trucks. (I say 'he' strictly through obedience to grammar. The phenomenon happens to women almost as much as to men. Under age 25, I'd say just as much.)

Back in his other life, back when he was urban or suburban, it may have been sports cars that caught the newcomer's eye. Or maybe a showroom full of compacts, fresh and glittering from the factory. Now he finds himself eyeing some neighbor's sturdy green pickup with a big load of brush in the back and wondering how much one like it would cost. Welcome to the club.

Pickups aren't necessary in the country, but they are certainly handy. Any rural family that can afford two vehicles should probably make one of them a truck. And it is quite possible to have a truck as a family's sole transportation. It is also quite economical, since pickups begin almost as cheap as the cheapest cars, and go up in price, size, and quality at the same rate cars do— except that about halfway up the car price range, you have reached the most expensive pickups there are.

What's handiest about pickups is their versatility. First, obviously, in load. Because of that big open space in back, you can carry almost anything. Two beef cattle. Two full-length sofas that you're donating to the rummage. All the apples in a small orchard. With the tailgate down, a load of sixteen-foot boards.

About 40 bales of hay. A full cord of firewood (provided it's dry). All your fence posts, your wire, and your tools, when you're building a fence. It's possible to get a little drunk with power, just thinking what a pickup can do.

Second, in range. Even without four-wheel drive, a pickup is a great deal freer than most cars to leave roads and drive over fields. Picking up hay bales, for example. Or to squeeze along homemade woods roads. This freedom comes partly because pickups are designed to have fairly high road clearance, even when loaded. Partly because they can takes tires that will walk you right through a wet spot or over a (not-too-big) rock. Most pickup owners in rural New England keep a pair of oversize snow tires mounted on the rear wheels all year round. And partly because you can shift weight around in a truck to get maximum traction in a way that would cause the average car to collapse on its fat Detroit springs.

Third, a pickup is versatile in function. Besides its truck role, a pickup can do anything a car can do, with one exception—about which, more later. It can drive you to work, for example, using no more gas than a car, and when you arrive, it will not only park in the regular lot, it will do so in a smaller space than a Cadillac or an Oldsmobile.

Nor are you going to complain on the way that it drives like a truck, because it doesn't. It drives like a car. The ride is reasonably smooth, the surreptitious U-turn reasonably easy. Drivers of big highway trucks have ten gears to shift, and air brakes to worry about. Drivers of pickups have a standard shift and regular car brakes. And if they hate shifting, most pickups can be had with automatic transmission. For that matter, most can be equipped with a stereo tape deck, so that you barrel out to the woods playing Beethoven. I admit that a fully loaded pickup–say, a Chevrolet C-10 with a ton of rocks in the back, or a six-barrel gathering tank full of maple sap–doesn't corner quite so neatly as an M G , but it still drives essentially like a car.

The one exception is that a pickup is not much good for carrying a big load of people. At least, not in the winter or in wet weather. On a sunny summer day, its capacity is something else. Giving hayrides at our local fair, I once had fifteen children and two mothers back there in the hay, plus myself and a friend in the cab. I make that eighteen passengers.

Even if two couples are going out to dinner, a pickup is not handy. Four adults will fit not too uncomfortably in the cab of an American (though not a Japanese) pickup–but the law says three. The other husband may or may not want to crouch in back. Furthermore, it doesn't take many bags of groceries to produce a sense of claustrophobia in a pickup cab. A mother taking two children shopping on a rainy day in a pickup generally wishes she had a car.

There *is* a solution, to be sure. People who go out to dinner a lot, or mothers with four children, can get a crew-cab pickup. This is not what you'd call a glamour vehicle. It has two cabs, one behind the other, and looks something like a centipede dragging a large box. But it does seat six people. The only problem is that you now have a truck not only so ugly but so big that it is no longer versatile in the woods. I do not recommend it.

One last advantage of pickups should be mentioned. They never make hideous noises or refuse to start because you haven't fastened the seat belt. Like other trucks, they are exempt from that law. You can use the belts when you're roaring down the highway and skip them when you're going one mile an hour in the woods. Handy.

So much for pickups and their virtues. The time has now come to discuss the art of buying one. It *is* an art, incidentally, unlike car buying. A few wrong decisions on options can cut a farm pickup's usefulness by 50%.

The first decision, of course, is new or used. A really old pickup, small and square and no-nonsense, is about the most charming vehicle there is. Also one of the cheapest. You can get one for $200. Any children you know will adore the running boards and–if you have an old enough one–the windshield that pushes open for ventilation.

On the other hand, old pickups tend to have unreliable brakes

31

and not notably reliable anything else. The 1947 Dodge I once owned—from its nineteenth to its twenty-first years—couldn't be counted on to start at any temperature much below freezing. That meant a long period of parking on hills and losing my temper, each October and November, until I finally gave up and put it in the barn until spring. One year I waited a week too late and had it frozen in the barnyard, in the way of practically everything, for three and a half months.

Even much newer pickups have generally led hard lives. (Little old ladies seldom own pickups.) Furthermore, it's difficult and expensive to have heavy-duty springs and other desirable country equipment installed in an existing truck. Probably only people with real mechanical ability should consider getting one.

But one last word. If you do get one, take a nice winter vacation and get it in some place like South Carolina. South Carolina pickups have never experienced road salt. At least the body won't rust out on you in a few years.

Now let's turn to new pickups. They come, basically, in three sizes and three styles. The sizes are called half-ton, three-quarter-ton, and one ton. Not one of these names means what it says. Which is a good thing, because a bunch of trucks that could carry only 1,000 to 2,000 pounds wouldn't be worth much.

Let me define the three. A half-ton is the basic pickup: what you find at a car dealer, what people mean when they speak of a pickup. At the moment it comes in two avatars. It is a small Japanese-made truck that can carry almost a ton of cargo. Or it is a somewhat larger American-made truck that, with proper springs and tires, can manage a ton and a half.

A three-quarter ton looks much the same, but has a much larger, truck-type rear axle. It costs more, gives a rougher ride, and carries loads of up to about three tons. People with campers put them on three-quarter-ton pickups—and then usually get about six miles to the gallon.

A one-ton has an even bigger rear axle. With dual rear wheels, it can carry up to near five tons. Neither it nor a three-quarter is what most people need for use on a country place. Not unless they plan to get into the lumber-delivery business, or maybe have always wanted, since they were kids, to have their own personal dump truck. (No fooling. There are truck shops in every New England state that will put a dump body on a three-quarter or a one-ton. The cost runs around $1,200. I have sometimes played with the notion.) But for general country use a half-ton is right; and for the rest of this article I shall talk about half-tons only.

Of the three styles, one can be dismissed right off. This is the tarted-up and chromed-up half-ton which attempts to pass itself off as a car. Chevrolet calls its El Camino; other makes have equally foolish names.

For families that are sincerely embarrassed at having to own a truck and that really think it would be preferable to drive something that looks like a scooped-out car, spending the extra money for an El Camino may make sense. Especially in those flat and treeless parts of the country where taking your truck out to the back forty must be something like driving across a very large football field. But to get one as a working truck on a rocky, hilly, wooded New England farm would be an act of insanity.

The other two styles are narrow-bed and wide-bed. Just as the half-ton is the classic American pickup, the narrow-bed is the classic half-ton. The design has been stable for 50 years now. Behind the cab you have–in most makes–a wooden-floored metal box four feet wide and six or eight feet long. (You get to pick.) This sits inside the rear wheels. Because it doesn't rust, and because it gives surer footing to any livestock you happen to be transporting, the wooden floor is a considerable advantage.

Until a few years ago, the narrow-bed was the cheapest of all pickups; now it costs exactly the same as a wide-bed. It retains two other advantages. Since the rear wheels don't stick up into the bed, you can slide slidable cargo in and out with great ease. And because this is the classic model, the tailgate in most makes is still the traditional kind that you hook with a chain on each side. That matters. You are able either to put such a tailgate down level, as an extension of the bed, if you are carrying a load

33

of long boards, or to let it drop all the way down for ease in loading. Now that I no longer have one, I miss it.

All the Japanese-made pickups and most of the current American ones are wide-bed. These have a cargo space five and a half feet wide (American) or four and a half feet (Japanese). Obviously you can carry a lot more cargo. On the other hand, the rear wheel housings stick in on each side, which is sometimes inconvenient. (I will say they are handy for children to sit on.) And on many wide-beds you get a fancy one-handed tailgate, like a station wagon's, which won't drop down unless you disconnect the hinges. It's not difficult, it's just tedious. And when you want to close the tailgate, you have to reconnect them.

Which style is better? Myself, I used to have a narrow-bed and now have a wide-bed. I think the advantages and disadvantages of the two models just about balance. So on my current truck, I made the choice on esthetic grounds. The narrow-bed lost. Properly designed, it is the truest of trucks, the very platonic essence of a truck. But in the last five years Ford, Dodge, Chevrolet, Jeep, and International Harvester (a G M C pickup is just a relabeled Chevrolet)–all have moved to such enormously wide cabs that a new narrow-bed looks hydrocephalic. Wide-bed is now the handsomer truck.

As to whether Toyota, Datsun, L U V, or an American make, the decision really rests on how much highway driving you're going to do. The Japanese trucks, with their four-cylinder engines, get far better gas mileage. According to *Consumer Reports*, they average about twenty mpg, while American half-tons average fifteen. My own experience suggests that Japanese pickups do a little better than twenty, and American pickups a little worse than fifteen. For a vehicle that I was going to commute to work in, and just use as a truck on the occasional weekend, I would probably choose to save gas (plus about $300 in purchase price) and get a Japanese.

But for a truck that was to be mainly or even considerably a working farm vehicle, I still prefer the larger and more versatile American pickup. It's not just that you can carry more weight, it's that you can get a specially adapted country model. The Japanese trucks tend to be unadaptable, the same for an appliance dealer in New York City as for a family with 60 acres of woods in Colebrook, Connecticut.

The man delivering refrigerators in the Bronx doesn't need any special traction. He never leaves the pavement. But the family in Colebrook does. And one of the most humiliating things that can happen is to get stuck in your own truck on your own place. Especially since you're so unlikely to be able to jack, or rock, or bull your way out.

The trick is not to get stuck. Which means that you may want the four-wheel drive available as an option (a very expensive one) on all the American but none of the little Japanese trucks. Otherwise, you will certainly want limited-slip differential. This inexpensive ($75 to $100) option means a special rear axle designed so that when one rear wheel starts to spin, all the power goes to the other wheel. Normally when one wheel starts to spin, all the power goes to *it*, and that's why you get stuck. Limited-slip differential is said to have its dangers, especially in very fast highway driving, where you may fishtail in a skid, but it is highly desirable in a farm pickup. I would rate it, quite impressionistically, as making about a third of the difference between regular two-wheel drive and four-wheel drive. You can get it on American pickups, but not Japanese. It's only fair to add that I know farmers with Datsun pickups who say they have no trouble at all zipping up and down their rolling fields, even on dewy mornings, but I still commend limited-slip differential.

If you decide on a Japanese truck anyway, about all you have to do is go get it. Maybe settle whether or not you want a radio. But if you opt for an American truck, you still have to pick your engine, with at least six more country options to consider.

The engine is easy. Get a six-cylinder. Almost all pickups—including three-quarter and one-tons—can be had either six or eight cylinders. A six has all the power you will ever need, and wastes less gas. As to options, the first and most important is to specify heavy-duty springs in the rear, and heavy-duty shock absorbers front and rear. All this costs about $40; its value in increased usefulness must be about 50 times that much.

Second, for any pickup that's going to leave roads, either a four-speed shift or automatic transmission is a great asset. Why? Because going through a field with long grass (and hidden rocks), or up into the woods, you need to be able to creep along, almost literally feeling your way, and still not lose momentum. Low speed in a three-speed shift will not let you go slowly enough.

Here the Japanese trucks have an advantage, since all of them come with a four-speed shift. It costs an extra $125 on an American pickup. But even better than four-speed is an automatic transmission. You can creep with astonishing slowness, and still have power. I have never owned a car with automatic transmission and never plan to, but on my sturdy green pickup I find it marvelous. It does, of course, use too much gas, and my next truck will be four-speed manual.

Third, you ought to get a step-and-tow bumper. Unlike cars, trucks are sold with no rear bumper at all. (How much rear bumper have you ever seen on a tractor-trailer or on a gasoline truck?) But a step-and-tow–which is a broad bumper covered with sheet steel–really is handy for pulling, and as a rear step. The ones you get factory installed, for about $50, are not nearly as sturdy as they look, but are still worth having. The ideal is to have one made by a local welder. Rodney Palmer, the man who owns the garage in Thetford Center, is a superb welder; and for $101.50 I have a rear bumper that will fend off anything short of a Centurion tank, that is heavy enough to give me good traction with no load in the truck, and that will last for a hundred years. (Rodney designed it so that I can move it from pickup to pickup for the rest of my life. Then I'll will it to my daughters.) Incidentally, if you don't get a bumper like that, you should plan to keep a couple of large flat rocks or about four cement blocks in the back each winter. Way back. Otherwise you'll find yourself spinning to a halt halfway up icy hills. Better anchor them too, so that if you have to slam the brakes on hard they won't come hurtling through the back of the cab and kill you.

Fourth, for a family truck it is worth getting extra padding in the seat. A pickup has a reasonably smooth ride, but not so smooth that additional cushioning won't be pleasing to visiting grandparents, people with bad backs, and so on. If you're going to get stereo tapes, you might even want to pad the whole cab, so as to reduce road noise.

Fifth, if you can talk the dealer into it, get him to remove the four automobile tires the truck comes equipped with, and have him put on four truck tires. They should be not merely heavier ply, but if possible an inch larger in diameter. And as I said earlier, the rear ones should probably be snow tires, even if you get the truck in May. (Come winter, put snow tires on the front, too. They won't improve traction unless you have four-wheel drive, but they will help astonishingly in preventing sideslipping.)

If you can't talk the dealer into it, you're no horsetrader. In that case, resign yourself to your helpless condition, and pay extra for big tires. Or go to another dealer. Or hurry home and read Faulkner's *The Hamlet*. Then you will learn—from a master—how to trade.

Sixth, get the truck undercoated. Presumably any New Englander knows about undercoating anyway—but it's even more important on pickups than cars, since people usually keep pickups longer. The process called Ziebarting is probably the best and certainly the most expensive. If you can stand having your truck smell like fish oil for a month or so, I recommend it. If not, a grease undercoating is said to be adequate. But the full mysteries of Duracoat (acrylic resin), asphalt, and all the other undercoatings, I do not pretend to be a master of.

There are all sorts of other machismo things one can get with a pickup. You can have a snowplow mounted—in which case be sure to get four-wheel drive. Plan also to have the front end realigned frequently, because plowing will spoil the wheel alignment with surprising speed. You can have an electric winch put on the front, and thus be sure of freeing yourself 99% of the time when you get stuck. (Though a two-ton manual winch of the kind called a come-along will do nearly as well. You can get one for about $45, and keep it under the seat.) You can have a power take-off on most larger pickups, and run your own saw-mill. You can merely buy a logging chain, keep that under the seat, too, and then when you find eight-foot poplars growing in the corners of your best field, you hook the chain on your step-and-tow bumper and pull them out by the roots. They don't grow back next year *that* way.

But just a basic pickup is machismo (or feminismo) enough. In fact, I can think of just one problem. Someday when you're going past the post office with a big load of brush, you'll glance up and see a whole row of summer people staring at you. With naked envy in their eyes.

Jan Lincklaen's Vermont

IT IS SEPTEMBER, 1791. A young Dutchman named Jan
Lincklaen is riding horseback up the muddy road from Rutland
to Burlington, Vermont. Once an officer in the Royal Dutch
Navy, Lincklaen is now in the real estate business. He is the
American scout for a giant land investment company back home
in Amsterdam. The company already owns four million acres
of land in New York State. Now it is thinking about buying
23,000 acres of maple sugar groves in Vermont. This is a pilot
agricultural project. If all goes well (it's not going to), Dutch
housewives will someday sweeten their coffee and frost their
cakes with Vermont maple sugar, instead of cane sugar from the
West Indies. Then the investors in Amsterdam can ease their
consciences. They won't have to feel guilty about owning slaves
to work the sugar plantations in Curacao and Aruba.

The road Jan Lincklaen is riding along passes through frontier
farm country. Most of it has been settled less than twenty years.
Vermont is still so wild that in the very year he makes his visit,
one upland farmer kills 27 bears in a six-month period. It is so
primitive that there is only one church bell in the entire state, way
over at Newbury. There are no school bells at all, and not too
many schools.

It is good farm country, though—boom country, like the Napa
Valley a hundred years later. Jan Lincklaen likes what he sees.
'The soil is very rich,' he reports, and adds that it is partic-

ularly good for growing wheat and Indian corn. Up near Burlington, farmers are getting 40 bushels of wheat to the acre, and up to 70 of corn. They are exporting beef to Canada.

There is wonderful hay land, too. Another visitor in the 1790's was dazzled to find farmers near Rutland who were making 500 tons of hay a year–20,000 bales, as we would say now. All this is from the virgin soil, which in another generation will be seriously depleted. Then farmers' sons from Vermont and New Hampshire will begin to stream out to the Middle West, to exploit a still richer soil.

But meanwhile northern New England is a place where a farmer can get rich. Nothing is commoner in those first 30 years of settlement, than to be able to buy a piece of forest for $1 an acre, clear it and get huge crops for a few years, and then sell out for ten or twenty times what you paid. Jan Lincklaen met a farmer in Dorset in 1791 who had just sold his 60-acre farm for $19.25 an acre. In terms of present money, that would be something like $500 an acre–not too much less than land in Dorset brings right now.

The earliest farms would have looked very ugly to modern eyes. When a new farmer arrived, his first act was to chop down every tree on what was going to be his first field, cutting them about two feet from the ground. (This is a good chopping height.) He would then cut these giant oaks and hickories and maples into lengths, drag them into piles with oxen, and burn them. Then he would gather the ashes and boil them into pot ash, later called potash. Pot ash was the basic ingredient in eighteenth-century soap, and he could sell it for a high price. Or he could refine it still further into pearl ash, which people used then for baking powder. 'These ashes amply pay them for the clearing of the land,' a Vermont lawyer named John Graham wrote in 1795.

Well they might. Vermont pioneer farmers were producing about two million pounds of pot ash and pearl ash a year, and getting the then enormous price of 3¢ to 5¢ a pound. Shipped by water to New York or Philadelphia, barrels of ashes from Vermont and New Hampshire sold for a higher price per pound than tobacco, or flour, or even butter.

The pioneer farmer now had a stretch of rich virgin soil, dotted at frequent intervals with enormous stumps. His next step was to build a house. Here is a contemporary Vermont account of how he did it. (I have added a little punctuation.)

'When any person fixes upon a settlement in this part of the Country, with the assistance of one or two others he immediately sets about felling trees proper for the purpose. These are one to two feet in diameter and forty feet or upwards in length...

'The largest four are placed in a square form, upon a solid foundation of stone. This done, the logs are rolled upon blocks, one above another, until the square becomes about twenty or twenty-five feet high. The rafters are then made for the roof, which is covered with the bark taken off the trees... The interstices in the body of the hut are filled up with mortar, made of the wild grass, chopped up and mixed with clay...

'In this manner is an abode finished, spacious enough to accommodate twelve or fifteen persons, and which often serves for as many years, till the lands are entirely cleared, and the settlers become sufficiently opulent to erect better houses. Three men will build one of these huts in six days.'

Looking out from his windowless, chimneyless house onto a landscape of stumps, the pioneer farmer did not see desolation. Instead he saw visions of a glorious future. Here on that knoll would go the ten-room clapboard house which he would start to build as soon as the new sawmill in the village got going. There on the bottom land he would grow his hemp (for making rope, not drugs—his highs came from life), his flax, his wheat. As soon as the roots rotted, out would come all those stumps. With the oxen he would drag them into rows, and fence off some grazing land for the cattle. Meanwhile, it was time to plant an orchard, and to begin on some stone walls.

In most of rural Vermont and New Hampshire, these dreams came true in a hurry. It was only 39 years from the settling of the first towns to the beginning of the nineteenth century. But men and women, three of whom can build a log house 40 feet square and two stories high in a week can create a whole landscape in twenty years; and by the year 1800 the two states looked pretty much the way they do now, in their unspoiled sections, except that the soil and the people were both richer in 1800.

Not that life was easy. The famous rigor of our climate was the same then as now. A man who spent some time in Newfane, Vermont, in the 1790's complained, 'This place is extremely cold and bleak in Winter, and not very hot in Summer.' There were wolves in the mountains in such numbers as to make the keeping of sheep almost impossible.

But there was also abundance and prosperity—and an arcadian simplicity that in our own day seems almost incredible.

While he was inspecting Vermont farmland, Jan Lincklaen paid a call on the biggest farmer and second most famous man in the state. This was Thomas Chittenden, Captain General and Governor of Vermont. He was then rich in years and honors, not to mention land. He had become Vermont's first governor thirteen years earlier, in 1778, and he had governed uninterruptedly for eleven years. Then he stepped down for a year—and when the young Dutchman came to call in 1791 had just been triumphantly restored to office.

There was no Secret Service detail, or even a state trooper. There was just an old farmer. He showed the visitors into his house 'without ceremony, in the country fashion.' Lincklaen, who was used to admirals and twenty-one gun salutes, could hardly believe his eyes. 'His house & way of living have nothing to distinguish them from those of any private individual, but he offers heartily a glass of Grog, potatoes, & bacon to anyone who wishes to come and see him.'

Maybe rural Vermont is a little like that still.

Selling Firewood in New York

ONCE A NEW YORKER, now a peasant, I live on a 100-acre farm in central Vermont. Now I'm driving back to the city with a load of firewood to sell. It's good dry maple, cut and split last year, plus I have about 50 gleaming white birch logs, cut last week and still oozing sap. You could hardly light them with a blowtorch, but they look pretty. I'm curious to see if New York wood dealers will buy them.

I'm curious about New York and firewood altogether. In Vermont, people buy wood by the cord—which is a stack four feet high, four feet wide, and eight feet long. Your average cord of firewood contains somewhere between 400 and 800 logs, and weighs a couple of tons. In Vermont it currently retails for about $55. Per log that would average out to 9¢ if we ever bought wood by the log, which we don't. But the myth is that city people do, and that they think nothing of paying $1.00 apiece. I mean to check that.

It's a beautiful October day as I roll down Interstate 91 with my load of logs. The one annoyance is that my pickup is getting lousy mileage with all that weight in back—maybe nine miles to the gallon. Halfway down, I have to stop for gas in the little city of Greenfield, Mass. The man at the pump peers over the sideboards. 'You taking that wood someplace to sell?' he asks. Yes, I say, New York.

'Jesus! you'll make a killing. Only you oughta have all birch.'

And he walks me over to the edge of the station. 'Look at that,' he says, pointing to a small pile of gleaming white logs on the next lot. 'Three for $1.00, that guy charges.' Golly. 'Yeah, he gets it up to his camp [this is New England peasant dialect for a summer cabin]. It don't cost him *nothing*. It's all profit.' The filling station man is actually turning green with envy as he talks.

So far so good. Only halfway there, and already wood is selling for 33¢ a log.

I could tell him something about getting it no-cost, though. Between a raw tree, so to speak, and a load of cut and split logs, there is an incredible amount of labor. You handle every damn stick five or six times. What does he handle? While I'm out in the woods swinging a nine-pound maul on hardwood logs with knots, he is sitting on his duff watching the gasoline truck pump his tanks full. But I don't get into that. I hand over $9.50 for the gas (*money* is what he handles), and head on down to the city.

Because we rustics always leave for market at dawn, it is just barely noon when I come onto the island and start rumbling down Second Avenue. I have a friend with me, and she is holding a list of wood dealers in her hand, copied weeks ago out of the Manhattan yellow pages. We are going first to Brusca Ice and Wood, at 95th and Third. There are eleven wood dealers on the island, incidentally, which is eleven more than there are in the Bronx. Brooklyn and Staten Island have three or four each, and there are two in Queens.

We get nowhere at Brusca, though I see at least five cords of nice-looking wood stacked up in the narrow little shop. Mr. Brusca speaks almost no English, and the *figlio* is out making a delivery. We never even get it across that we are trying to sell wood, not buy it. (How come he doesn't guess this from my truck? Because having been a hayseed for fourteen years now, I don't have the nerve to double-park, and my truck is a block and a half away.)

We also get nowhere at A.A. Armato, Second and 88th. Nowhere at National Ice Service, on East 80th. In fact, we don't even get a look at National's wood. What's at the address turns out to be an answering service. Six surprised-looking telephone operators turn their heads simultaneously when we open the door to their tiny windowless office.

So we head for Clark & Wilkins of Park Avenue. This is not the Park Avenue that elegant New Yorkers mean when they speak of Park Avenue. This is up at 127th Street. The 'avenue' here is two narrow canyons, as completely separated by the great stone railroad embankment in the middle as two Vermont towns are by the mountain rearing up between them. Clark & Wilkins is a hundred-year-old white firm in what is otherwise a solidly black neighborhood. Also the biggest wood dealer in the city. I didn't go there first only because I wanted to practice up before trying the big time.

My friend has to stay in the truck, because this time I double-park right out front. Half-forgotten city ways are coming back. Clark & Wilkins is a large brick building, sort of fortress-like. Inside, it is dim and quiet. Several men are cutting four-foot bolts of wood into short pieces and then tying them up in elegant little bundles. All are neatly wrapped in burlap.

After a minute the foreman gives me the eye, and I explain that I have a lot of good dry Vermont maple just outside. Instantly he shoos me out. He says pleasantly but firmly that they have a regular supplier who provides the top-quality, super-dry wood that New Yorkers demand.

Since he didn't give me time to ask his prices, and I want to know what those bundles cost, I now send my friend in to pose as a customer. She is from New Hampshire, but to my mind she could pass as a lower Park Avenue type any day. She also knows wood. What she finds is that they sell *all* their wood in burlap bags (doormen hate bark in the lobby), and that all the bags cost $5.00. You can get a bag with twenty pieces of miniature firewood twelve inches long, or a bag with fifteen pieces one foot, four inches long, or a bag with twelve pieces a foot and a half long. She checks a few out, and finds they really do contain dry beech and oak, with a very occasional piece of punky elm.

46

Selling Firewood

While she is doing her detective work, a Clark & Wilkins employee named Rocco comes out for a cigarette, and spots my truck with me skulking behind it. Then he sees the white birch, and this turns out to be the one thing their regular supplier doesn't have. Two minutes later the foreman is also outside, and we're dickering. I hadn't planned to mention either that my birch is green or that some of it isn't white birch at all, but an inferior species called gray birch. But as we dicker, the foreman casually reaches out to heft a piece—and seeing that green wood weighs just about twice as much as dry, I go honest and tell him I cut it last week. He shrugs. 'So? The customers won't care—what do they know? Birch they just use for decoration, anyway.'

Honesty satisfied, I ask him to make me an offer for the whole 50 logs. He looks me straight in the eye. 'My friend, I have absolutely no idea what it's worth. [In Vermont we usually look at our feet when saying something so outrageous.] You give me a price.'

So I do. I ask for $10.00—20¢ a log—which would be robbery in Vermont, but seems reasonable here. Apparently not reasonable enough, though. He looks me in the eye again, and says that the big boss really does all the buying, and he's not in. We agree I'll come back tomorrow.

But I want to sell my wood today. I'm scared that if I don't it will be stolen tonight. There is no way to lock the back of a pickup truck, and I am probably going to have to park on the street. This is because New York garages are violently prejudiced against trucks. The only other time I ever came to the city in my pickup, I got run out of four parking garages in a row, and wound up in an open-air lot about a mile from where I was staying. I hope to manage better this trip. My friend and I are staying tonight at the Algonquin, and the manager of that hotel has kindly agreed to try to find a garage that will take my truck. It is, after all, no bigger than most cars—smaller than a Cadillac. But I don't dare count on his succeeding, and meanwhile it is already after four.

So we roar down from 127th Street, looking for more wood dealers. It isn't easy. Manhattan wood dealers are inconspicuous.

No vulgar signs or advertising. The headquarters of Cassamassima & Sons on East 50th, for example, turns out to be a basement apartment in an old brownstone. When Mrs. Cassamassima comes to the door in a housecoat, I feel sure I have the wrong place. But no, it's the right one, only her *sposo* is off making a delivery. Would he like some nice dry maple? No, he's overstocked. Birch? No demand for it.

Finally, just as it's getting dark, we come to Diamond Ice Cube, at 201 East 33rd, and here our luck turns. They are still open, and they're buying wood. The boss looks my truckload over, and concedes that the logs on top are suitable. 'Of course, I don't know what you've got under there,' he says suspiciously. (More dry maple, that's what.) How much do I want for it, he asks.

It's almost six o'clock, and I am eager to sell that wood. I tell him I'll take $30 for the entire five-eighths of a cord of maple. The birch, too? No, the birch would be extra. 'Thirty for the lot,' he says firmly. 'Thirty-five,' I answer, equally firmly, and he agrees. This is the same old 9¢ a log, not worth the trip down, but at least I've got a sale.

What I don't have is a delivery. Six o'clock is quitting time, the boss says, and his drivers have gone home. He wants me to come back tomorrow to unload. I explain that the wood will probably be stolen by then. He explains that that's my worry, not his. But luck is still with us. The Algonquin has done nobly, and by six-thirty I am crawling up the ramp of the Kinney Garage on West 44th with my ton of wood. It is noticed immediately.

There are three attendants, all black. The one who comes over to the truck is a young guy, very lively. He has a piece of white birch in his hand in no time, and is stripping off a piece of bark. 'Hey, man, it's just like paper.' I tell him the Indians used to write on it.

'This tree don't grow in America, do it?' he asks in honest surprise. I assure him I cut it myself in Vermont, but he isn't really listening. He's noticing the difference between the white birch (chalky, and it peels) and the gray birch (not chalky, doesn't peel), missed by three wood dealers earlier. I tell him

48

this, causing the oldest black attendant, who is now also standing by the truck with a piece of wood in his hand, to jeer, 'He don't know the difference between that tree and his rump.'

The young guy ignores that, too. 'Do it carve?' he asks me, still holding the log. 'I'd like to buy me this.' The encounter has been too pleasant to mar with a 9¢ sale; I give him the log. 'Thanks, man,' he says. 'I'll whittle me something nice.'

At the Algonquin I change from work pants and a blue denim jacket to slightly oudated city clothes, and my friend and I proceed to blow most of the wood money ($21.50, to be exact) on dinner at a Spanish restaurant. The food is not as good as I remember its having been the last time I was there, fourteen years ago. Then, like proper hayseeds, we go to Radio City Music Hall.

The next morning we are back at Diamond Ice Cube at nine sharp. The load is intact, down to the log the old guy picked up, which I recognize because it is still lying at an angle where he tossed it back in.

Two drivers are waiting, and the four of us unload my wood and pitch it into a cellarway. As we do, we talk, and I learn quite a lot about the wood business in New York. I learn, for example, that it's been in a slow decline for about 75 years, as more and more buildings with fireplaces are torn down. But thanks to the energy crisis, there is probably going to be an upturn, starting this year. I learn that all eleven Manhattan wood dealers use the bag system; prices range from $3 to $6.50 a bag. That is, from about $175 to $375 a cord. Diamond itself charges $5 a bag, just like Clark & Wilkins. I learn that Diamond's biggest cost, apart from the wood itself and the burlap bags (they have to pay 25¢ each for them) is the parking tickets they get while delivering. They budget for two a week, at $25 each. I learn that wood is delivered in such small quantities because of doormen. Doormen like neat, clean packages that don't weigh much. In fact, what they really prefer are the artificial logs made in California out of wax and sawdust (Diamond sells these, too), which come wrapped and sealed like loaves of bread. Fortunately, most lower Park Avenue types want real wood. So do photographers, who constitute a small but steady market for birch. They buy three logs at a time, and use them in decorator shots.

But the thing I learn that surprises me most has nothing to do with firewood in New York. One of the two drivers turns out to be from St. Albans, Vermont—he moved to the city when he was sixteen—and I am telling him what a rip-off the Spanish restaurant was, and how much cheaper things still are in Vermont. He smiles pityingly. 'Twenty-one fifty is *nothing*,' he says. And he tells me that last winter he took his wife and two children skiing at Bolton Valley, in northern Vermont, and got through $900 in four days. My friend asks him if a lot of this wasn't for rental of equipment, and he says no, he and his wife have all the gear. Skis for the kids, yes, they rented them.

I am still absorbing this vision of Vermont as more expensive than New York when the boss comes back from a delivery. He inspects my wood, likes it, writes out a check, and tells me he'll take as much more as I care to bring down. My friend and I head uptown in the empty truck, hit a couple of museums—among other things, we hear a free lecture by the Countess of Jellicoe on 'The Bath in Art' at the Frick—and then start back north. The next day, which is a Sunday, I cut and split a full cord of white birch, and stack it in the barn to dry. I'll be down with it next year. I want another one of those cheap weekends in the city.

A Cool Morning in Vermont

IT'S A RARE VERMONT WINTER that doesn't have one stretch of weather when it's twenty below every night, and not much above zero even at midday. Keeping warm during such a spell is either difficult or expensive–sometimes both. If you're living in a big old house in the country, and if you have an oil furnace, you've got roughly three choices. You can keep the thermostat up at 70 and go broke. You can turn it down to 55, put your family in long underwear, and shiver. Or you can heat two or three rooms with wood stoves, and move in, relying on the furnace only to keep the pipes in the rest of the house from freezing. That's what most of us do.

Last winter, though, my old farmhouse was getting a long-promised remodeling. The little wood stove that normally sits in the corner of the kitchen and the big one that normally dominates the living room were both out in the barn, surrounded by dismantled stovepipe. My wife and daughters had gone on a trip for the three weeks we'd be without a kitchen. I was alone in the house, doing the fifty-five-and-shiver routine.

One specially cold morning I woke up with an uneasy feeling. The house was deathly silent. No distant cellar hum of the furnace, no comforting purr in the hot-air ducts. Pausing only to put on my long underwear (and a wool shirt, and wool pants, and a sweater, and heavy socks, and boots), I ran downstairs. We have a thermometer in the living room. Thirty-seven in

there. We have another on the front porch. Twenty-six below zero out there. I figured I had maybe two hours before the house began to freeze.

I hurried down cellar and began pushing every start and reset button I could find on the furnace. Silence as before. I circled the furnace again, found a small red button on the left side, and jammed it hard. The blower sprang to life. I ran upstairs. A chill wind was pouring through the ducts. Not a trace of heat. And ten minutes gone already.

Back down to poke at the furnace some more. Then I suddenly remembered the oil tank. I made it across the cellar in two jumps, and looked at the gauge. Empty.

While I was on the phone to the oil company, the church clock in the village (half a mile away) struck eight. As it struck, Heman Durkee and his son Heman, Jr., the two carpenters who were remodeling the kitchen, arrived. They were just in time to hear my cry of anguish when I learned that the delivery truck had already left on its morning rounds. The manager would try to catch it at its next stop, but even if he did, it would be 30 miles away. He couldn't promise what time it would get to me, except to assure me that it would be in the forenoon.

Heman is temperamentally a stoic. 'Guess we can work just as well in our coats,' was all he said to me as he and his son stood watching their breath form in the kitchen. But Heman, Jr., frankly prefers creature comforts. 'Didn't you have a little stove in here?' he asked.

Ten minutes later we had the stove set up on the sub-flooring, the pipe run up, and a good fire of carpenter scraps going. Then the three of us staggered in with the big old parlor stove, and set that up. It's a great gothic-looking stove made in Rutland about 100 years ago, and it will take two-foot logs up to about nine inches in diameter. Before I had to leave for work at quarter to nine, the living room was practically warm, and a few wisps of heat were even floating upstairs toward the two bathrooms and all their tender pipes.

Heman called me at work around eleven to say that the house

was safe. The oil truck had come and put 258 gallons into my 250 gallon tank. Though he is not only stoic but taciturn, he added something else. 'Guess you're going to save a little money,' he said.

'How is that?'

'Don't they give you the first 50 gallons free, when they let you run out?'

I hadn't known that. But it's true. They do. Whether they would have insisted if I hadn't brought the matter up, thanks to Heman, is another question.

In fact, I saved more than that. Heman and Heman, Jr., kept the little kitchen stove going every day, until the actual morning came to put the new floor down. By then, the cold spell was over, and we were having a thaw.

They never said it was as a favor to me. Or to keep them warm. Perhaps it wasn't. 'That much less scrap to take to the dump,' Heman explained.

Market Research
in the General Store

ABOUT FIFTEEN MONTHS AGO, the Best Foods Company made a big splash with a series of TV ads about pancake syrup. They filmed the ads in Vermont. In them, native after native was shown tucking into two samples of pancake, and then saying that he or she preferred the one soaked in a Best Foods product called Golden Griddle to the one soaked in Vermont maple syrup.

These ads upset a lot of people in Vermont, including the Attorney General, who got an injunction against them. They also upset me. Because if Golden Griddle was really better, why was I working so hard every spring? Why was I hanging sap buckets, gathering, boiling, falling into snow drifts, when I could just as well be down working at a Best Foods factory in New Jersey? So I decided to check this matter out.

The first thing I did was to buy a bottle of Golden Griddle. That is, I bought a plastic container filled with a mixture of sugar, dextrose syrup, corn syrup, sodium benzoate, potassium sorbate, natural and artificial flavors, and caramel coloring. Plus 3% maple syrup. The next Sunday morning I had my wife serve me two identical pancakes, just as in the TV ads. One was covered with Golden Griddle and one with my own maple syrup.

I had no trouble telling them apart. The Golden Griddle had a nice color, and it's certainly sweet enough. But it had quite a perceptible chemical taste. I voted the maple syrup first by a wide margin. Then I gave the test to her and our daughters. Same results.

This was such fun that the following Sunday morning we decided to do it again. We invited two couples to breakfast, old friends who happen to be fellow sugarers. Four more votes for maple syrup. Maple syrup now ahead eight to nothing.

By now it looked pretty suspicious. How come all the Vermonters in the TV ads like Golden Griddle better, and all the ones we tried like maple syrup better? But before deciding that Best Foods was pulling a fast one, we decided to wait one more Sunday. This time we invited some other old friends—father, mother, and two teenage children—who must be something like eighth and ninth generation Vermonters. We were not expecting the result. Three votes for Golden Griddle, one for maple syrup.

The three who had picked Golden Griddle were pretty unhappy about it, being good Vermonters. Naturally we spent the rest of breakfast discussing what made them choose it, and doing more tasting. All three finally decided it was because Golden Griddle has such a strong flavor.

At this point I developed a theory. As anyone who has read this far knows, maple syrup comes in three grades called Fancy, A, and B, with progressively stronger tastes. There's also a fourth variety, stronger still, which never appears in retail stores. Officially, it's called 'ungraded syrup,' but locally everyone calls it Grade C.

I had been using Fancy at the three Sunday breakfasts. What if I had used B or C? Would the father and the two kids in that family have still preferred Golden Griddle? I decided I would run a much bigger test, this time using two kinds of maple syrup as well as Golden Griddle. But first I would find out exactly what Best Foods had done, so I could compare my results to theirs as accurately as possible.

After quite a lot of writing and phoning, I learned that they had hired a New York market research company called Decisions Center, Inc., to come to Vermont and do the whole thing. Decisions Center had done a good and careful job. They spent three days testing 223 people, of whom 58% had preferred Golden Griddle, 40% had chosen maple syrup, and 2% hadn't been able to decide. And, sure enough, they had used a mild Grade A syrup from a big producer down in Windsor County. I know him. The only possibly sneaky thing in the whole operation was done by Best Foods itself, not the market researchers. The tests were given in a shopping center in the little industrial city of Spring-field, Vermont. The people tested were naturally mostly from Springfield. But when it came time to make the folksy commercials, all these city people were taken over to Newfane, a picture-postcard village, so the background would look more rural and authentic. But that's probably normal advertising technique.

The first free day I had, I hustled down to Windsor County and bought a quart of the identical Grade A syrup they had used. Then I opened an old mayonnaise jar of my own Grade C, picked up the bottle of Golden Griddle, and set off for the shopping center in my town. That is, I walked over to the general store. All one morning, my daughter Amy and I sat at a table in the Village Store in Thetford Center and ran tests. Forty people tried our three samples. That represents everyone who came into the store that morning, except a few on diets and two who have diabetes.

What we found was fascinating. About a quarter of the human race have naturally good palates—or, at any rate, a quarter of the people in our test did. That is, about a quarter of the people we tested not only had a preference, but could identify the different syrups by taste. After the test they'd say Sample 1 is early-run maple, Sample 2 ain't maple syrup at all, and Sample 3 must be end of the season. They were right.

All nine people who could identify the samples put the Grade A first. All but one of them put the Golden Griddle last. They hated it. Elmer Brown, for example, who runs Brown's Nursery, and is a native of northern Vermont. The minute he tasted Sample 2, he looked at me accusingly and said, 'Why, Noel, that one's got Karo in it.'

Of the other 31, three liked the Grade A best, thirteen liked Grade C best, and fifteen liked Golden Griddle best. So my test results are as follows. One hundred per cent of the people with good palates preferred maple syrup, and so did 52% of the people without good palates.

But I don't see any great surprise in that. Why wouldn't they? Maple syrup is free from potassium sorbate and sodium benzoate; it has absolutely no synthetic smell. The important finding, as I see it, is that 90.3% of the people with untrained palates wanted a powerful flavor. Something really strong. What I interpret this to mean is that if you're a gourmet, it's well worth getting Fancy or Grade A maple syrup. Maybe even if you just *want* to be one. But if you're not and don't care whether you ever are, you're wasting your money. Being a maple producer, I am hardly going to suggest that you therefore get Golden Griddle or Log Cabin or Vermont Maid (which seems actually to be made in Winston-Salem, North Carolina). Instead I suggest you get a good hearty Grade B, and save $2 a gallon. Or since you won't find it in any store, write to some farmer and get a gallon of Grade C direct from him. It currently costs around $9 a gallon, which isn't all that much more than the supermarket stuff. Golden Griddle, Log Cabin, etc., if you ever bought a whole gallon at once, would run you between $6.50 and $7.50.

You don't know a farmer to write? I know lots, and I have an obliging publisher. If you write me care of him, I will undertake to pass orders on. For Fancy, A, B, or C. After all, I'm not plugging my own syrup. What with those tests and my regular customers, not to mention making 300 maple sugar hearts for our village fair last summer,* I'm sold out.

* I made them out of the rest of my Grade C. Personally, I wouldn't dream of putting C on pancakes. As syrup, I only like Fancy, A, and B.

Buying a Chainsaw

IF I WERE TO MOVE to an old-fashioned farm, everything quaint and hand made like a scene from Old Sturbridge Village, and could bring just one piece of modern machinery with me, I wouldn't hesitate a second. I'd bring my chainsaw. It's noisy, it's dangerous, it pollutes the air—and I love it.

Not every rural New Englander would make the same choice, I admit. A dairy farmer, used to 80 Holsteins in the barn, would be almost certain to bring his milking machine. Someone seriously into beef cattle, used to mowing 150 acres of hay a summer, would doubtless bring his tractor. But for us amateur farmers, and for people who are living on a country place without doing any farming at all, the obvious choice would almost always be a chainsaw.

The reason is simple. Nearly all of New England is naturally forest. Leave a field alone for ten years, and it comes up trees. Popple and wild cherry race in from one side, gray birch and pine from the other. Leave a trail you cleared in the woods alone for about two years, and like magic the trees and brush on both sides have reassumed possession. Men who want clear spaces have to keep cutting.

Not only that, what they cut has become, rather abruptly, a valuable product again. The day of fossil fuel is almost over. The day of cheap nuclear fuel is not yet. (And may never be.) Mean-

while, the casual trees—yes, even the dead elms—a New Englander cuts around his place are worth a good deal of money.

For many years, firewood has been a middle-class luxury, something people used because open fires are pretty. But in the last four years it has turned into a standard heating fuel, the most economical in New England. Provided, of course, you use it in Franklin stoves, box stoves, Defiants, Jøtuls, etc., and not in fireplaces.

For possible skeptics, let me take a second to prove this. A cord of dry hardwood produces roughly the same amount of heat as 190 gallons of fuel oil. No two cords are the same, of course. A cord of rock maple yields more B T U's than a cord of elm; and even two cords of elm will differ, depending on where the trees grew, what the proportion of limb wood to trunk wood is, maybe even on what time of year you did the cutting. But one cord per 190 gallons remains a good average figure.

When fuel oil sold for 18¢ or 19¢ a gallon, which it did around here until 1973, a cord of wood therefore yielded $35 worth of heat, measured in terms of oil. But it was likely to *cost* $40 or $45. Each stick had to be hand fed into the fire. Luxury stuff. But last winter oil cost 51¢ a gallon here, which made the same wood worth $97 in terms of the heat it yielded. Meanwhile, a cord was selling for $55. A bargain! And with the right kind of stove or wood furnace, you need only stoke once every few hours. I run three stoves these days—two full-time, and one in the guest room when we have guests. My wife and I have also done some fierce insulating. Our oil bill is slightly lower than it was ten years ago.

One can, of course, clear land and cut firewood without a chainsaw. They certainly did in Sturbridge, when it was New Sturbridge. I myself got along my first eight years in the country with a Swedish pulp saw and an axe. In those days my wood-cutting was quieter, safer, and cheaper in terms of equipment than it is now. But it also was about five times as slow, which matters when one is making a serious woodpile. (I would say a serious woodpile begins at two cords.) Furthermore, my work lacked a kind of deftness it now has. Never again will I voluntarily live in rural New England without at least one chainsaw. I'd rather have, and do have, two—a heavy-duty one for regular work and a light one for pruning.

People who set out to buy a chainsaw for the first time usually make two mistakes. The first is that they tend to get one that's too small—sometimes too small even for much pruning work. The major manufacturers—McCulloch, Homelite, Baird-Poulan, and the German company Stihl—each put out a cute little saw that sells for somewhere between $99.50 and $149.50. It's light, it's easy to handle, it seems just the thing for an amateur. It isn't. Or, rather, it isn't unless you expect to use your saw no more than about an hour a week. Almost none of them will stand up to the work of getting in a winter's wood supply. They may be all right for playing lumberjack in a suburban yard—though hard on the neighbors' ears, since most of them scrimp on the muffler. And even for casual work these toy-poodle saws are really good only for toy-poodle trees. I mean those up to ten inches or a foot in diameter. You *can* fell and buck up a two-foot oak with one, and it will be easier than gnawing it with your teeth, or than chopping it, unless you're a really fine axeman, but not much. The chainsaw dealers I know say that there was a run on mini-saws when the energy crisis hit—followed by a run on considerably larger ones, as one by one the early buyers discovered their mistake.

What most country dwellers are going to find meets their needs is a medium-sized saw that sells for about $250. The cutting part, the bar and chain, won't be much bigger than those of the toy poodles—sixteen or eighteen inches long, rather than twelve or fourteen inches. The main differences will be in the larger engine, the better-built sprocket, starter assembly and so on, all of which will give you a much faster-cutting and more reliable saw. You'll also have a little more weight to lug into the woods.

Saws of all sizes come with a choice of automatic or manual oiling. I have tried both. I find the automatic slightly more convenient, and also slightly more wasteful of oil. In the end, I opt for manual, but I don't think it's an important choice. The only important thing is never to find yourself, as I once did, owning one saw of each kind. Your thumb will never remember whether to press the oiling button or not.

Some saws also come with an optional self-sharpening device, at considerable extra cost. I have never tried one. But I do notice that you can buy three or four new chains for the price of the sharpener. Anyway, you can get a file and file-holder for not much, and sharpen the saw yourself on rainy days. Preferably in a vise. If I can, anyone can.

There are big saws, of course, as well as medium size. The McCulloch 105, for example, weighs 22.3 pounds, as opposed to six or seven for a toy poodle and ten to fourteen for a medium. You can get a four-foot bar on it, if you want to (though twenty or 24 inches would be the more usual choices). If you're going into the logging business, it's something to consider. Otherwise it makes about as much sense as getting a ten-ton truck for use around the place.

The second mistake that most new buyers of chainsaws make is one they may not discover for fifteen or twenty years, when their hearing starts to go. A chainsaw is noisy to the point of danger, a fact not much stressed in those T V ads where music plays all the time the man in the ad is merrily cutting. All chain-saw users need to wear ear protection.

Except in California, Oregon, and Washington, where state law requires the use of ear protectors, very few chainsaw dealers ever mention the subject. Some–I know two–do stock acoustical ear muffs or plugs, and will sell you a pair if you know enough to ask. To be fair to the rest, many dealers simply don't know the danger, and (since they are test-running saws every day) will themselves be going deaf in a decade or two.

They don't know because the manufacturers don't tell them. With the distinguished exception of Stihl and some of the Swedes –whose frankness may be explained partly by the fact that their saws are the quietest on the market–most manufacturers tend to deny the problem. A spokesman for McCulloch, for example, says that 'it is extremely doubtful that any hearing impairment will occur' from chainsaw use, and that certainly there is no need for a moderate user like myself to worry about protection.

It's a different story when you talk to acoustical engineers or professional lumbermen, however. A big lumber company like Weyerhauser requires its employees to wear ear plugs on the job–and also encourages them to take the plugs home if they are going to do their own cutting on a weekend. Acoustical engineers say that possible hearing damage begins at 90 decibels. Most Poulans, Homelites, and McCullochs are in the 110 decibel range, most Stihls and Swedish saws in the 105 range. Because decibels are figured logarithmically, that's a difference in sound level, not of 5%, but of 25%. Get protection. If you don't care whether you look funny or not, get muffs. If you do, get plugs.

There is one last question. What brand? I have just one piece of advice. And that's not to get a house brand, such as Sears Roebuck. At least, not unless you know and trust the local store. The saw will be O K (it'll just be a Homelite or whatever with a different nameplate), but repairs and service will be a pain in the neck. Myself, I tend to get whatever brand has the best serviceman near my place. At the moment, that gives me a choice of McCulloch, Homelite, and Husgvarna, handled by Dave Fitzgerald of Norwich, Vermont. If Dave were selling live beavers instead of chainsaws, I'd probably stick with him. But I'm glad he prefers chainsaws.

In Search of the Perfect Fence Post

MIDWEST FARMERS—most of them, anyway—have a boring time with fence posts. When they need some more, they just open the Sears catalogue and order another 500 metal ones. (Current price: $2.49 each.) Then they lay out 100 or so in a straight line across the prairie, and start stringing wire.

New England is different. Within a mile of my house—well, two miles—I can look at fences strung on eleven different kinds of posts. Bud Palmer, who runs the garage in the village, has a horse pasture fenced with pine. Ellis Paige uses mostly split oak to restrain his Angus cattle. Barbara Duncan keeps her goats behind a mixture of oak and maple saplings. George deNagy uses hemlock for Push and Pull, his team of work ponies. Warren DeMont has metal posts. Not from Sears, but salvaged from a floodplain the Government took over a few years ago. Floyd Dexter, the best fencer of us all, uses entirely sharpened cedar posts. Except when he runs out in the middle of a job, that is, and then he's been known to use cherry, tamarack, lever wood …almost anything except popple or elm. The town uses cut granite posts, six inches square and six feet long, as a good many farmers used to. My neighbor Dr. Lucius Nye has—but let me get on to my own story.

I've been fencing for sixteen years now. My serious fences surround three cow pastures totaling 32 acres. The posts are mostly cedar and butternut, decently soaked in oil.

But I also have a *jeu d'esprit:* a little half-acre sheep pasture (for two sheep) done entirely in green hemlock. Five young apple trees fenced against deer with a variety of posts that could be called New England Miscellaneous. Plus more. Over the years I've probably put up two dozen fences of one sort or another. And since I started from a state of ignorance which farmers' sons usually pass beyond between the ages of five and six, I have made every mistake but one that it's possible to make. I've put up a fence without bracing the corners. Strung barbwire from the bottom strand up instead of the top strand down. Put the small end of a post in the ground instead of the large. The one thing I *haven't* done is to use white or gray birch for posts. And there it was poetry that saved me, not common sense. Long before I thought of being a farmer, I had read most of Robert Frost, and could quote from 'Home Burial':

> *Three foggy mornings and one rainy day*
> *Will rot the best birch fence a man can build.*

It's not even much of an exaggeration.

Sixteen years ago I blundered into fence building when I acquired a wife and an old farm the same year. She was determined to have a garden, and the deer were determined she wasn't. I volunteered to build a fence.

Having all that newly acquired land, most of it covered with trees, I wasn't about to buy posts. I took my pulp saw and wandered around my new woods until I found a place where there were a lot of gray-barked young trees coming up three and four together. I know now they were sprout-growth elms and red maples. Even then I knew it was good forestry practice to take a tree coming up in three or four stems like that and cut it back to one main stem. So I sawed busily until I had 20 nice gray-barked posts, enough to fence the garden, six posts to a side. (If you think that would take 24 posts, you clearly haven't done much fencing. And if you retort that it would be 24 after all, because a big garden needs two gates and two posts for each, you'd be right. But I hadn't thought of that yet.)

The next step was to get them into the ground. Someone had told me that fence posts should be set two feet deep, so I went to one corner of the garden and dug as narrow a two-foot hole as my shovel would dig. Then I put one of the four largest posts in, shoveled the dirt back, and tamped it. Then I tested my new corner post, which easily moved through a twenty-degree arc. I spent the next half hour packing around it with rocks, and then retamping with another fence post rather than with my foot. The post now wiggled only slightly.

It was absurd to think of spending 40 minutes per post, so I did what I usually do when I'm in a dilemma. I went to Dan and Whit's (in the opinion of many, one of the world's great general stores) to see what exotic tool they carried that would solve my problem.

Almost immediately I found a thing called a post-hole digger, and brought it home in triumph. It worked splendidly. The holes it made were a little too big for my posts, but only a little, and I could get fairly tight posts planted in about ten minutes each. No deer got into the garden that year.

Three years later, however, they could have shouldered their way in almost anywhere. Elm posts rot fast. Green elm posts rot faster. Green elm posts with the bark on rot fastest of all. If I were ever mad enough to use elm posts again, I would cut them in the spring when the bark is loosest, peel them, dry them for a year, and then soak them in used motor oil.

But I never have used them again. Not counting a small grazing area for pigs (ineptly electrified, and a total failure), my next fencing project was to make a paddock for a horse we had bought. This was the year that the first eleven garden posts rotted out; and overreacting as usual, I had decided to use posts that would damn well *never* rot out. In fact, I got metal posts from Sears. The date was 1965, and back then the posts were a mere $1.29 each.

I am the sort of person who has three little helpings at dinner instead of one giant one, and this temperamental quirk carries over to buying farm equipment. Thus, though I figured the paddock would take two rolls of barbwire and 70 or 80 posts, what

I actually bought was one roll and 30 posts. This was good luck, because I quickly discovered that however handy metal posts are on the great plains, they aren't much use on the rollercoaster terrain of a New England farm. You ask what about Warren DeMont and his salvaged metal posts? He uses them on stock fence for *his* two sheep a year. Stock fence goes on relatively level ground, and they work fine. I'm talking about barbwire up and down hill. You can't drive staples into a metal post, and it's nearly impossible to get the wire tight.

So I switched in mid-fence to cedar. By this time, in my fourth year with land of my own, I knew several of the farmers in town and had walked land with them. I knew about cedar posts, and about driving mauls. I had also learned to recognize most of the common trees that grow in New England, and I knew that I personally had no cedars at all. No matter. They sell them at Agway. In fact, I might have gone to Agway even if the farm had been one big cedar swamp. The garden fence and the pig fence had been such failures that I'm not sure I would have trusted a cedar I had cut myself. I wanted something professional.

There were two gigantic piles of posts out behind the main Agway building in White River Junction. All cedar, all sharpened at the big end to a beautiful tapering point. Posts from one pile cost $55 a hundred (Agway is getting $95 for the same posts now), and posts from the other pile cost $45 a hundred. I am sorry to report I did it again. I got posts from the $45 pile. It was my last really *major* mistake in fencing.

Though any normal person enjoys saving ten dollars, it wasn't for that that I got the cheap posts. I got them because they were smaller. They ran 3″ to 4″ at the butt end, while the $55 posts were 4″ and up. I committed this idiocy out of fear. While I was vainly trying to string wire on the metal posts, a farmer friend had given me a couple of sharpened cedar posts so I could see what I was missing. One was nearly 6″ in diameter (he said use it for a corner post), and one was quite small. Even with my new twelve-pound driving maul from Dan and Whit's, and my wife to hold, it had taken me about 40 full-strength blows to drive the big one. I had to stop in the middle and rest. The last six or eight blows pretty well splintered the top, and pretty well ruined me. By contrast, the smaller one had gone into the ground 2" a whack, and left me feeling like the strong man at a county fair. So I concluded that big driving posts were for real farmers, and little ones were for transplanted urbanites like me.

This was, of course, a completely false conclusion. For which I paid the price six years later, when I had to redo the whole paddock because all my tiny posts were giving out. The true conclusion is this. You have to learn *how* to drive sharpened posts, and when you do, it's easy. You don't even need someone to hold the posts up for you (though it's more companionable that way).

Here is the true way to put up posts for a New England fence, learned from sixteen years of hard experience. Or, rather, here are two ways: the fast way and the best way.

The fast way is to go buy however many posts you need. Make sure you buy large ones, 4" to 5" in diameter at the butt, and sharpened so that the taper extends at least a foot. Throughout New England, such posts are normally cedar–though if you find locust, grab them. Meanwhile buy or borrow a soft iron driving maul–not to be confused with a sledge hammer, which has a much smaller head–and a good four-foot iron bar. Lay out your first strand of wire to make a line, or else do it with string. Then you can start driving posts. Take your bar, and drive it into the ground where you want the first post to go. If you hit a small stone, you can drive it on down with the bar. If you hit a big one, or ledge, move.

When you are down about twenty inches or so (usually about four easy whacks with the maul), stop. Then wiggle the bar around with a circular motion until you have a hole sort of like the inside of an ice cream cone, with the top about 4" in diameter. Then pull the bar out, pace off the distance to the next post, and stick the bar into the ground where the post is to go. (That way you won't lose it.) Now go back and shove a post firmly in the first hole. It will easily stand, and you should be able to drive it in anywhere from six to a dozen blows, depending on what the soil is like. And it will be completely tight–no play at all.

When you get to the big corner posts, you can do them the same way–but it won't be any six to twelve blows. If you have a post hole digger, now is when it's useful. It will make you a straight-sided hole about 5″ in diameter. Dig one a foot deep. Then work your bar for another foot in the bottom. Plump the 6″ corner post in, and drive it with ease.

That's all there is to the fast way. Provided, of course, you remember to brace all corner posts with braces that go right to the base of the nearest line post on either side. Little bitty braces do nothing. You can buy cedar poles for braces, or you can cut your own. Either way, they should be about a foot longer than the distance between your fence posts, and at least 3″ in diameter at the small end. Here's how to install them. With a shovel you make a small hole right up against a line post, and shove the butt end of the brace in. Then you lay the smaller end carefully on top of the corner post, and trim it to length with a neat diagonal cut. Then you slide it down the corner post for about a foot, which gets it good and tight, and nail it in place with a tenpenny (3″) nail. Rich people sometimes use sixteen or even twenty-penny nails.

The best way is considerably more complicated, and also takes more equipment. But it's cheaper, and Robinson Crusoe would like it better.

First learn to recognize all the trees you have. If you don't already know how to chainsaw, learn to. Then start looking for stands of young trees that need thinning. In the absence of cedar, wild cherry or tamarack is best, though both hemlock and white pine will do. Don't bother with trees growing in the open; they taper too fast, and you'll get only one or two posts from a tree. Three to five is what you should be getting. Since you cut the posts six feet long, that means the tree should have eighteen to 30 feet of straight trunk before it gets too small. Cut a bunch.

Another option is to split out posts. Butternut probably works best. If you had some butternut trees nine to fifteen inches in diameter, growing straight and not too many lower limbs, you can make a lot of posts from one tree. You fell it, buck it into six-foot logs, and split each log into fourths or sixths. All it takes is two wedges, a sledge hammer, and a modicum of skill. Butternut is a notoriously weak wood, so you always make large posts. I myself wouldn't dream of sacrificing a good butternut tree just for posts–but where I wanted to thin a fenceline anyway, or where a butternut was growing too close to sugar maples I wanted to favor, I occasionally take one. I've probably made a hundred posts that way. Six trees' worth.

If you're smart, you will have cut all these posts where you can get pretty close to them with a pickup truck, which you now drive out there. Bring your wife (or husband, or unsuspecting houseguest) and an extra pair of ear protectors. Open the tailgate and load the first three or four posts in the back of the truck, with the butt end sticking out. The spouse or guest puts on the ear protectors and climbs in the back of the truck. While he or she holds the first post steady, you sharpen it with your chain saw. This amounts to cutting a slice off each side the full length of the chainsaw blade, getting the victim in the back of the truck to turn the post 90 degrees, and then cutting off two more slices. The whole procedure takes less than a minute. It leaves, incidentally, a pile of fluffy shavings like giant excelsior, which children find irresistible. Now do the other two or three, have the victim pile the sharpened posts on one side, and load the next batch.

If you're *really* smart, you will have done all this in the spring; and as each post is sharpened, you can also peel it. The four bark points where the sharpening cuts end will pull like Band-aids. I myself am rarely that smart and, to be honest, I think it matters only moderately.

By lunchtime you will have posts enough for a great deal of fence; and if you want to you can start pounding that afternoon. I have. If you're the deferred-pleasure type, though, instead bring them home and pile them under cover for a year. And the next spring, if you have the time, treat them. With a few posts, you just paint the bottom two and a half feet with creosote or whatever. But if you've cut a lot, that gets exceedingly boring– and besides, painting doesn't give deep penetration. The better way is to give them a 24-hour soak. I have usually done this in an old 55-gallon drum, in which I put a mixture of one-third creosote and two-thirds old motor oil. Say, five gallons of creosote and ten of motor oil. Three-thirds creosote would doubtless be better, but creosote is expensive. You can soak about fifteen posts at a time. I strongly advise tying the barrel to a tree, because when it's full of six-foot posts, it is top-heavy; and few things are more annoying than having a barrel filled with used motor oil and creosote tip over on top of you, or even not on top of you. The smell, my wife informs me, does not come out of work pants until the third washing.

Having done all this, you are ready to take your iron bar and your driving maul, and set posts. You will have a double satisfaction when you're done. You can look at your new fence and reflect that thanks to your skill it's going to last two and perhaps three times as long as ordinary fences. And (provided you take care not to count the purchase price of any mauls or chainsaws—which, after all, you still have, and will keep using), you can compute that your posts cost you about 10¢ each. That's chiefly for the creosote.

On the other hand, part of me hopes that you don't do all this—that you go and cut some basswood in the morning, and have it in the ground that afternoon. True, it will rot out as fast as my elm fence around the garden. But after three years it may be time to move a garden, anyway. And what else are young basswoods good for? Besides, you'll be contributing to that sense of variety which I hope New England never loses.

There are two Yale dropouts who are caretaking/renting a house about three miles from our village. They've made a big vegetable garden, and they have fenced it with chickenwire mounted entirely on alders. Skinny alders driven small end into the ground. What's more, it works. A deer could probably push that fence over with one hoof, but they don't. Nor do they jump over. I think they're as pleased and touched by that fence as I am, and wouldn't hurt it for the world.

Raising Sheep

THE DIFFERENCE BETWEEN 'a place in the country' and a farm is chiefly a matter of livestock. It is in New England, anyway. You can own 200 acres, you can pick your own apples, you can buy a small tractor—and you're still just a suburbanite with an unusually large lot. But put one cow in your pasture, raise a couple of sheep, even buy a pig, and instantly your place becomes a farm.

Since most people who move to the country sooner or later do get the impulse to try farming—or if they don't their children get it for them—I want to talk about livestock. Pruning apple trees and buying tractors can come later.

I shall make two assumptions. The first is that most people who read this have little or no experience with farm animals. You not only don't know how to geld a pig or butcher a steer (after a dozen pigs and four steers, I don't either), you've never so much as raised hens. The second is that at this moment you are un-decided what animals to keep.

Right now is a good time to make a decision—especially if 'now' is fall, winter, or early spring. If it's summer, you're going to have to wait nine months to get started; if it's late spring, you have rush preparations to make. A farm animal should be out and grazing by the first of May, which means it should be bought in April. And meanwhile a new owner has considerable fencing to do, or he has a pen to build, or at least tethers to prepare.

The easiest animals to keep are sheep. The best animals to keep are beef cattle. The most interesting animals to keep are pigs. My advice is to try all three—but not the same year.

Your first year, start with either sheep or cattle. Two sheep or two cattle. They need each other for company. Don't bother with purebred stock, just get whatever happens to be available where you live. Any breed of sheep will do; Hereford cattle are probably the best for beginners, as being the gentlest and the least likely to jump fences. If you have extra time and energy, you can throw in a couple of pigs, as well. Again, any breed, though so-called bacon pigs are preferable to so-called lard pigs.

Then the second year you can try sheep if you started with cattle, or try cattle if you started with sheep. Goats? Forget them. Goats, except milk goats, are a folly. And if you're going to be milking, you might as well go all the way and get a milk cow. You're equally trapped at home either way. (Though I have to admit that a nanny provides the right amount of milk for a small family—and there *are* people who prefer the flavor of goats' milk. I also admit that goats are handy for clearing the brush in an abandoned pasture—provided they stay in it. If they get out, a thing goats are good at, they also clear the lilacs from in front of the house, the bark off sugar maples, anything their wicked little teeth can reach.)

My own decision, my first year, was to start with two lambs. There were several reasons. First, unless you're going to winter your stock, you have to begin with year-old cattle. A this year's calf is still veal when fall comes. Two yearling cattle need at least two acres of fenced pasture, with a water supply. And there is *no* easy way to fence two acres. It takes three-fifths of a mile of barbwire and 120 fence posts for a three-strand barbed fence, or a fifth of a mile of smooth wire and the same number of posts for an electric fence. But lambs you can tether for the few months until they turn into lamb chops and roast leg with mint jelly.

Second, you can bring two lambs home in the back seat of a car, whereas even a pickup truck will just barely handle two yearling cattle. The lambs will still fit in the car when they go to be slaughtered in the fall. Meanwhile, it is no great matter to move a lamb from one place to another, or to catch one that escapes. Driving cattle is harder. My second year, with Herefords, before I learned that a fence good enough for horses is calves' play for Herefords, the pair of them got out four times. Three of those times I was late for work. The last time was the Fourth of July. The day began with a 6 A M phone call from a neighbor who was understandably cross at having two large red-and-white cattle in her flowers. It ended at dark, by which time my wife and children were long gone to the fireworks in Orford, New Hampshire. Robert Frost never said good fences don't make good neighbors when there are cattle involved.

The last two reasons are less relevant. One was that two lambs seemed to me safer company for my then very small daughters than two large beef cattle. I have since learned that Herefords and little girls have a natural affinity. I am speaking, of course, of polled—which means hornless—Herefords, and even among them, of heifers and steers. Gentle as the breed is, I don't know that I'd want little girls playing in the pasture with a bull. The other reason was the high cost of lamb chops, which I adore. Now that beef is almost equally expensive, this is no longer much of an argument.

Anyway, let's assume that you decide to start with lambs. The first thing is to locate a couple. You can buy them at a stock auction, such as the one held every Monday evening in East Thetford, Vermont. But you don't get a 90-day or even a one-day guarantee, and people have been known to sell sick lambs. Better save auctions, exciting though they are and despite the bargains you can sometimes get, until you know something about sheep.

Another way is to get a subscription to a state farm bulletin. The two I know best are New Hampshire *Market Reports* ($4 a year from the New Hampshire Department of Agriculture in Concord) and the *Agriview,* free to Vermont farmers, amateur and otherwise, from the Vermont Department of Agriculture, Montpelier. Both are stuffed with ads for horses, cattle, sheep, pigs, and even lowly goats. So far as I know, every state has a comparable publication.

But the best way is to find a sheep farmer within ten miles of you. Since a sheep herd consists basically of ewes, most farmers are willing to sell an occasional male lamb, which they will probably call a buck. And most of them are willing to be telephoned when you have problems. Besides, it's just generally a good idea to get to know some real farmers when you set out to be an amateur one.

As for cost, it varies considerably. If you're willing to get lambs so young that you'll have to feed them with a bottle for a while, you can sometimes still find them for $10 or $15 each. A weaned lamb is likely to be at least $25 or $30–it would be $40 or $45 if you were foolish enough to get ewe lambs.

Besides the actual two lambs, the only equipment you need this first year is two tethers and two collars. The tethers can be no more than two fifteen-foot pieces of clothesline, tied to two stakes set 30 feet apart. (Any closer and you'll have two tangled lambs within half an hour.)

I do not recommend this, however. As summer goes on, you'll be moving the lambs every few days, and it gets harder and harder to drive the stakes deep enough. Furthermore, lambs tend to be extremely non-intellectual and are almost sure to get the rope completely wound around the stake, no matter how cleverly you notch and loose-tie. Much better are the light metal chains already attached by a rotating swivel to a corkscrew stake that you can buy at farmers' supply stores. The large size, which is the only one worth considering, costs around $3.

The collars should be leather, and as broad as possible. If you use a homemade rope or wire collar, the lambs are sooner or later going to chafe their necks, and maybe even break the skin. In that case, you'll be buying ether to kill the maggots that hatch in the wound, which is no treat at all.

You now have a mobile field-maintenance unit. In the course of the summer, two lambs can mow, trim, and fertilize just about half an acre of pasture. All the owner has to do, besides notice how much they enhance the view, is to move them every few days and keep them watered. Once a day is plenty, since they get most of their water from the grass. In the hottest part of summer, it is a kindness to tether them where they can get some shade, but they don't *have* to have shade, as pigs do.

And for people who don't even want to do the moving, there is an alternative system. Instead of tethering lambs, you can anchor them. That is, you fasten your rope or chain to a medium-sized rock, and let the lambs move themselves. Or since a lamb will wind his rope just as readily around a rock as a stake, you can get one of the metal corkscrews and pour a cement anchor around it. About 30 pounds will do. I admit that this system works better for one lamb than two; two will gradually tug themselves closer and closer until they are thoroughly entangled.

Along in late October, your lambs are ready to go on the dinner table. By this time you will know whether sheep are your thing. If they are—if you mean to go on with sheep next year—you might want to shear them, just to find out if you can do it. Preferably advised by a neighbor who grew up on a farm and has done it before. Even from lambs, and even sharing the wool with him, you will have enough left for a couple of natural-wool sweaters.

It's already getting cold in late October, and unless you plan to take the lambs into the house for the night, you should slaughter and butcher on the same day you shear. If you do it yourself, you first make sure there are no children around, or at least no children who realize what's about to happen. Then you shoot each lamb in the head, and immediately afterwards cut its throat. The carotid artery is what you are after. When the blood is out (it takes less than a minute), you dress them off.

If this sounds brutal, horrible, and ghoulish, don't even consider trying it. Take them, in the back seat of your car, to the nearest country slaughterhouse. You can probably get them back as cuts of meat already wrapped for the freezer. Then it's time to start planning for next year.

If you're going to go on with sheep, you'll want an actual small flock next year. Maybe four ewes, already bred when you buy them. You can spend the winter deciding whether you want Suffolks (a particularly elegant breed, with dark heads and legs—

sort of the Siamese cats of the sheep world), Dorsets, that fine meat breed, Hampshires, or whatever. Meanwhile, before the ground freezes you had better start fencing two or three acres with woven-wire sheep fence. Cattle fence won't hold sheep.

On the other hand, if you're going to try beef cattle, which I hope you are, now is the time to locate 120 cedar posts for *their* fence. You might as well get a driving maul at the same time.

Flow Gently, Sweet Maple

LAST YEAR was one of the great syrup years of this century. I have never seen maples run harder than they did on March 11th, 1977, and again on March 21st, and again on March 28th. Only once or twice have I ever known the sap as sweet. Even the poorest trees I tapped were running sweet to the taste. The best trees were behaving as if they thought they were sugarcane.

By the time the season finally ended, Vermont had produced 437,000 gallons of syrup. No other state produced even nearly as much. New York, five times as big as Vermont, came the closest to providing competition. New York produced a piddling 320,000 gallons. Many Vermont sugarers, myself included, made double what they did in 1976.

Sounds nice, doesn't it? Sounds like the kind of year you'd like to have every year. Well, it's not. I can barely stand it the one year in every ten or twelve that it does happen. Listen to the story of the wonder-year 1977, and then you'll understand why I hope we don't have another like it until about 1990 . . . or maybe 1995.

The season began quietly. There was no hint of the excitement to come. I started hanging buckets on March 2nd, which was a day of bright sun and cold wind after a good hard freeze. Just like normal. By March 8th, I had my whole 104 buckets up—twenty on General Miller's lawn in the village, about 50 scattered around my own land, and the rest up the hill at Alice

Lacey's. The next day, which was a Sunday, I placidly made my first gallon of syrup.

At this point it still seemed a typical year. There had been a couple of pretty good days, with three or four inches of sap in good buckets. (Half an inch in bad ones.) There had also been several days when nothing ran at all. But then on March 11th there was a quantum jump. We suddenly got a day when it was probably warmer in Vermont than in Miami Beach. The temperature rose from 26 at 7 A M to 70 at noon. It clearly excited the maples. By 10 A M some spouts were actually running a tiny stream, instead of just dripping the way they are supposed to. I gathered all afternoon. By suppertime the holding tank at the sugarhouse was full, and the gathering tank on the back of my truck was nearly full. We had just one worry. The sun was so powerful that the sap in some buckets was warm to the touch, and we feared it would spoil before we could make syrup.

So I hurried. Over the next two days, boiling late after work, I made ten gallons of syrup—which in my small evaporator takes thirteen or fourteen hours. No sap spoiled, but the grade of the syrup gradually dropped from Fancy down to A, and then on the last couple of gallons on down to B. Meanwhile, the weather stayed warm. We went three nights in a row without a frost. Then it rained for two days. Practically all the snow melted, and people began to worry that the season was already over, after that one splendid run.

They needn't have. On March 20th it snowed again, and on the 21st the maples went wild. Trees that normally drip a scanty half-gallon on a good day were filling a sixteen-quart pail right up. Two hours after I got home from work I had the holding tank full, and the gathering tank full, and 30 buckets still to gather. Most of them, full to the brim, were briskly running good sap onto the ground.

To make room to empty them, we'd have to boil. My wife announced an indefinite delay in supper, and fired up. By dark she had used enough sap so that I could empty each of the last 30 buckets halfway. Then I put them back on the trees, each with two gallons of sap still in it. This is not easy even by daylight. It's downright hard when you're doing it in the dark by feel.

The next day was a Saturday, and I boiled from 7:45 A M until 6 P M, making eight and a half gallons, all Fancy. But I didn't gain an inch. The trees ran almost as hard as they had yesterday. In the afternoon, while I furiously boiled, my wife and daughters gathered for me, getting all but eight buckets they didn't know about, which are on a little hill behind the sheep pasture.

I might have tried to go on boiling by candlelight after it got dark, but just before 6 P M I ran out of wood. Don't sneer. My normal year's production is 25 gallons, and that's what I keep wood enough at the sugarhouse for, with a small margin for safety. This year I had already made 29 gallons. Instead, after dinner I went out with a flashlight to get those last eight buckets. The radio said we were going to have a blizzard with twelve to eighteen inches of snow. It would be a lot easier to climb the hill to those buckets now than tomorrow. Big wet snowflakes were coming down hard as I hurried up the hill at 9 P M with my flashlight and gathering pails. All eight buckets were running over, and I had four round trips to make, stumbling in the dark. (The flashlight helps in emptying, but you put it in your pocket when you're coming down with a full gathering pail in each hand.)

Sunday is a day of rest. For people who don't sugar, that is. Me, I got up at 6 A M to resume boiling. I noticed with joy that the blizzard hadn't come. There were barely three inches of new snow. By 6:30 I had loaded my truck with firewood from the barn (good hardwood from the house supply, not the pine slabs and spruce tops I normally use at the sugarhouse), and was off. On my way I checked a couple of roadside buckets, and my spirits immediately sank. The trees had been running all night. Both buckets were half full.

Worse news was in store at the sugarhouse. Much worse. When I ran out of wood the evening before, I had grabbed a few oak logs from the barn to finish off the last gallon I made. And being very tired, I had also failed to check the float valves

8 3

when I quit boiling. One of them, I now discovered, was jammed. No sap had come into the finishing pan all night.

Normally that wouldn't have mattered; pine and spruce leave no coals, and the pans would have cooled off before any harm was done. But oak is another matter. The pans had been gently steaming all night—indeed, the sap pan still was. As for the finishing pan, it held a twisted mass of black carbon—all that was left of what would have been two gallons of Fancy syrup. It took three hours to scrape and scrub it clean. I tried to be grateful that at least I hadn't ruined the pan itself, just melted a little solder.

Meanwhile the sun came out and the maples, which had now been running 24 hours straight, picked up speed. By the time I could start boiling at 10 A M , some buckets we had emptied yesterday afternoon were two-thirds full again. Both tanks, of course, were completely full.

The next day was the low point of the season. Though it was a Monday, and though I was bone-tired, I stayed home to boil. I didn't want to keep on using dry housewood—I didn't have that much left. I took just enough down to get the fire going well, and then began feeding in some red maple I'd cut just a couple of months earlier.

The fire practically went out. I could make the pans simmer, but except when I borrowed more dry wood from the barn, I couldn't make them boil. Between 7 A M and 3 P M I managed to produce only three and a half gallons. Then I gave up, and went to the woods to cut dead elms. Standing dead elm is the only dry wood you *can* harvest in March. I naturally also checked a couple of buckets, and noted with sorrow that though it was a bleak windy day when by rights the trees shouldn't be running at all, they were in fact dripping briskly along. It was no hard run—but my 104 buckets would probably yield 40 or 50 gallons today, and me with no storage space, and no wood to boil with.

I took two more half-days off from work that week. Gradually I worked my tanks down, and built the supply of dead elm up. But during the enormous run on Friday, I lost every bit I had gained. Friday at dark I had 44½ gallons of syrup, full tanks, and no more elm.

Most people in town who make syrup were in a comparable plight, and at this point some of them pulled their buckets and called it a season. One of them, in fact, tried to give me three barrels of good sap he had on hand when he quit. I eagerly refused it.

I didn't pull my buckets, though. Not that I wasn't tired of sugaring (and cutting dead elms), but having tapped the maples, I felt it would be a sort of betrayal to remove the spouts and let the sap just dribble down the bark. A little as if a worker at a Red Cross blood center came up to a donor and pulled the needle out of his arm after half a pint, explaining that they'd filled the day's quota, and suggesting he just go bleed on the street. I couldn't do it. I kept boiling.

When I boiled the last time on April 6th, and closed down, I had made 57 gallons: 14 of Fancy, 23 of A, 12 of B, and 8 of C. I had prettied up something like five acres of woodland by removing all the dead elms. I still had the evaporator and all the buckets to wash. And I was so tired I slept twelve hours a night for the next three nights.

This spring I won't mind if we have a good season, but if we have another spectacular one, I shall probably quit midway and take a trip to South Carolina. Might even get a pickup while I'm there.

The Two Faces of Vermont

WHEN YOU CROSS THE BRIDGE from Lebanon, New Hampshire, to Hartford, Vermont, practically the first thing you see on the Vermont side is a large green and white sign. This bears two messages of almost equal prominence. The top one says, 'Welcome to Vermont, Last Stand of the Yankees.' The bottom one says, 'Hartford Chamber of Commerce.'

Only Vermont could have a sign like that, I think. Vermont makes a business of last stands. Consider just a few. It is the last stand of teams of horses that drag tanks of maple sap through the frosty snow. It is the last stand of farmers who plow with oxen and do the chores by lantern light. Together with New Hampshire and maybe a few places in Ohio, it is the last stand of dirt roads that people really live on, and of covered bridges that really bear traffic. It is the last stand of old-timers who lay up stone walls by hand, of weathered red barns with shingle roofs, of axmen who can cut a cord of stovewood in a morning— of, in short, a whole ancient and very appealing kind of rural life. This life is so appealing, in fact, that people will pay good money to see it being lived, which is where the trouble begins. There's a conflict of interest here.

On the one hand, it's to the interest of everyone in the tourist trade to keep Vermont (their motels, ski resorts, chambers of commerce, etc., excepted) as old-fashioned as possible. After all, it's weathered red barns with shingle roofs the tourists want to photograph, not concrete-block barns with sheet aluminum on top. Ideally, from the tourist point of view, there should be a man and two boys inside, milking by hand, not a lot of milking machinery pumping directly into a bulk tank. Out back, someone should be turning a grindstone to sharpen an ax–making a last stand, so to speak, against the chainsaw.

On the other hand, the average farmer can hardly wait to modernize. He wants a bulk tank, a couple of arc lights, an automated silo, and a new aluminum roof. Or in a sense he wants these things. Actually, he may like last-stand farming as well as any tourist does, but he can't make a living at it. In my town it's often said that a generation ago a man could raise and educate three children on fifteen cows and still put a little money in the bank. Now his son can just barely keep going with 40 cows. With fifteen cows, hand-milking was possible, and conceivably even economic; with 40 you need all the machinery you can get. But the tourists don't want to hear it clank.

The result of this dilemma is that the public image of Vermont and its private reality seem to be rapidly diverging. My favorite example comes, of course, from the maple-sugar business. Suppose you buy a quart of syrup in the village store in South Strafford. It comes in a can with brightly colored pictures on it. These pictures show men carrying sap pails on yokes, sugarhouses with great stacks of logs outside, teams of horses, and all the rest. They are distinctly last-stand pictures.

But suppose you decide to go into the sugaring business for yourself. When you write away for advice, you get a go-modern or private-reality answer. You are told not to hang pails at all, much less carry them to the sugarhouse on a yoke. Instead, install pipes. Don't bother to cut any four-foot logs, you're told, even though your hills are covered with trees. Texas oil gives a better-controlled heat. And finally, your instructions say, the right way to market the stuff is to put it in cans that show men carrying sap pails, sugarhouses with great stacks of logs....

The state is full of this kind of thing. I've seen a storekeeper spend half an hour taking crackers out of plastic-sealed boxes and putting them in the barrel he thinks summer vistors expect him to have. I've driven over a fine old covered bridge, intact and complete from floor to roof, and just as busy with modern traffic as it ever was with wagons. Tourists stop constantly to get pictures. But should one of them go poking around underneath (it doesn't happen often), he would see that it secretly rests on new steel I-beams, set in concrete. The great wooden trusses up above are just decoration now.

Or take fairs. I've been at a fair where the oxen for the ox-pull were trucked in from as far as fifty miles away. The town was full of oxen. If you didn't happen to notice them arriving in the trucks, you'd have concluded that here was a real last-stand neighborhood. Or you would have until the pulling began. Then you might have gotten suspicious. What the teams were pulling was more concrete, big slabs of it. Furthermore, when each pair of oxen had made its lunge, a distinctly modern element appeared. This was a large yellow backhoe which would rumble up, belching diesel fumes, and give the slab a quick push back to the starting point. The net effect was rather like watching the Dartmouth crew at practice, which I've also done. The college boys are like the imported oxen. They use muscle-power. The crew surges up the river between Vermont and New Hampshire, every man pulling his oar for dear life. The coach is like the backhoe. He skims alongside in a fast motor-boat, steering casually with one hand, and shouting orders through a megaphone he holds in the other.

Most of all, though, I see the difference between Vermont in photographs, in sentimental essays, in advertisements, and the state as it is actually getting to be. I'm thinking, for example, of roads. Even in California they know what a Vermont road is like. It's a last-stand road. It may be dirt or it may be blacktop, but what matters is that it's narrow and it follows the lay of the land. In most of Vermont, obviously, that means going in curves. The road will curve in so as not to spoil a field, curve out again afterwards, meander up a hill. It has, of course, a stone wall running along each side. Generally a row of big old trees marches beside each wall. Often these are maples, and when they are, the farmer who owns them taps every spring, using buckets.

But what if some Californian gets sick of twelve-lane expressways and moves to Vermont? What if he buys a house on such a road? He hardly gets the place remodeled (exterior unchanged, interior restored to authentic 1820, cellar packed with shiny new machinery) before the town road commissioner comes to see him.

The town's going to resurface the road next summer, the commissioner says. While the crew are at it, they plan to make a few other changes. They're going to take out all the sharp curves, reduce all the steep gradients, and widen the whole road by six feet. Twenty feet, if you count shoulders.

To the Californian's horror, it turns out that this will mean taking all the stone walls on one side, and most of the trees on both sides. It also turns out that the road will no longer follow the lay of the land. In particular, it's going to be raised four feet where it passes his house, and the road commissioner is hoping to use his stone wall for part of the fill. Next year the photographers will have to find some other road to put on their 'Unspoiled Vermont' calendar. But two cars will now be able to pass in mid-winter without one stopping and the other slowing down to ten miles an hour. And reality and image will be a little further apart.

If the ex-Californian puts up a fight for his stones and his trees, he soon finds that the selectmen and the road commissioner are not wholly against him. They may think he shows his Los Angeles background in wanting to save a stone wall when it's barbed wire you need for keeping cows, but they don't really disapprove. In fact, the road commissioner freely admits to liking last-stand roads himself. He was raised on one. What's on everybody's mind, it turns out, is that the town is not going to get any State Aid unless it widens and straightens the road to state specifications. And, of course, a lot of people in town are tired of having to stop every winter when they see another car coming. But the money is the main thing. The commissioner rather thinks the state itself gets Federal road money on similar conditions. In other words, town and state are under the same pressure all the dairy farmers are: go modern or go broke. That's a strong pressure.

And yet it's not the only one. Opposed to it is the natural cussedness of Vermonters, lots of whom don't want to go modern. And some would say it's not just cussedness, either. There are deep satisfactions to last-stand life. And, finally, there is all that good money the tourists pay.

All this has amounted to almost equal pressures in the two directions, at least until very recently. Almost everyone in Vermont is at least partly on both sides. But most are more on one side than the other. By oversimplifying a little, one can draw up a sort of chart of the battle lines. In fact, I have.

Let me start inside the fort. Manning the loopholes, and actually making the last stand of the Yankees, are a hard core of hill farmers, country storekeepers, ox breeders, and so forth. Economically their pressure is small. Most of them earn less money every year. But they aren't about to quit. In my part of the state, a fair number have taken full-time jobs so that they can keep farming nights and weekends. These are the kind referred to on the sign.

Allied with them are about half the summer people. (The other half aren't opposed; they're neutral. In fact, they're mostly too busy water-skiing and playing golf even to have noticed that there *are* farmers in Vermont.) But the first half like coming to a region of old-fashioned farms, and having farmers for neighbors. They may not want to look after cows or lay up stone walls themselves, but they like to watch other people do it. Meanwhile, the money they pay out for care-taking, barn-painting, and meadow-mowing helps to keep a good many last-stand families going.

Also allied are nearly all the middle-class immigrants or so-called year-round summer people. Most of them were originally drawn here by last-stand life, and a certain number actually lead it. I know one couple, both with college degrees, whose first action on getting their Vermont farm was to disconnect the electricity. They do the chores by lantern light. I know another man, born and bred in Maryland, who has become as good a country plumber and as authentic a rural character as lives in New England.

Finally, there is a scattering of people outside the state who provide economic support in one way or another. Here are the covered-bridge lovers who send money to help a Vermont town keep one. The bridge I mentioned a while back drew contributions from no less than four covered-bridge clubs last year when the town it's in had to decide whether to repair it or to replace it with the latest thing in concrete bridges. Here also are the city people who will spend extra time and money to get locally made cheese, or barnyard eggs from hens raised organically, or hand-made wooden toys—and in doing so have put a good many country stores in the mail-order business. If you could only get that by mail, too, some of them would buy hill cider that's capable of turning hard, rather than the tame stuff (filtered, pasteurized, and practically castrated) that's available in supermarkets. If they only had trucks, some of the suburban ones would to come up and buy half a ton of real manure for their gardens. The number of such people is small but growing.

Turning to the other side, an equally mixed group is pushing toward modernization. In the center are what I guess to be a majority of all native Vermonters under fifty, starting with the valley farmers who already have big herds and bulk tanks. They don't want to be the last stand of the Yankees. (After all, look where Custer was after *his* last stand.) They want their sons to be able to go on farming after them—even if the 'farm' turns out to be a lot of hydroponic tanks inside a two-acre concrete shed, fronting on a twelve-lane expressway.

Nearly everyone concerned with either education or state government is also on this side, at least officially. So are all of us who drive to shopping centers instead of walking to the village store, who buy lumber at Grossman's instead of at the town sawmill. And so, with a superb irony, are many Vermonters in the tourist trade, plus the tourists themselves.

The irony is that the tourists don't know they are. They come here to look at last-stand life. They wouldn't cross the road to look at a supermarket or a two-acre concrete shed. Most of them firmly believe they're helping to support old-fashioned Vermont by coming here at all. But though they flock to see the last-stand country, and, if they're here in the spring, to take a free taste of hot maple syrup, or in the fall to do a little free hunting—free as far as the owner of the land is concerned: the state charges a stiff fee—inevitably where they spend most of their money is in the motels, filling stations, and restaurants. Last-standers get only a little directly. They don't get much indirectly, either. Even though the restaurant owner knows that his tourist customers have come to look at last-stand life, and even though he personally hopes it will survive, he's still in business. He mostly buys his eggs at the battery farm, his milk at the big automated dairy, his beef from Kansas City, and so on. His chief gesture toward last-standism is to make sure the syrup cans in his gift shop have pictures of sap buckets on them.

In the last five years the balance has perceptibly tipped in favor of modernization. Most people agree that the last stand is likely to end in about one more generation. What will happen then? Let me present an admittedly partisan view.

Most of Vermont will look like—well, it will look like central New Jersey with hills. Where there are now fields and meadows, there'll be scrub woods mixed with frequent tree plantations. Every now and then there'll be an automated concrete 'farm.' Around each lake will be a ranch-style summer resort. The entire state will be linked by superb highways. In the more rugged sections, these highways will take most of the valley land there is. (Right now a four-lane highway built to Federal interstate specifications consumes 40 acres out of every square mile it goes through, or one-sixteenth of the whole square mile.)

There will, to be sure, be three or four villages left in which last-stand life goes on. Two of these, I guess, will be commercial ventures, and two will be owned by the state. All four will be pure fake. If you drove into one—I'm going to call it Old Newfane Village—first you'd see a wooden barn with four live cows in it, and a man specially trained to milk them. Then you'd notice a grove of maples next to an old-fashioned sugarhouse. Probably the maples will have to be made of plastic, with electric pumps inside, since the main tourist season begins in June rather than March, and since there's no way to keep a real maple from budding until June. But it will be genuine maple sap that the electric pumps draw up from a refrigerated tank under the sugarhouse.

Beyond the plastic maple grove will be a large woodshed. There, for 50¢ you'll be able to watch a man first sharpen his ax on a hand-turned grindstone and then chop up a couple of logs. Every twenty minutes he'll reblunt his ax by smacking it into a block of granite. An expert from Colonial Williamsburg will check his technique twice a day. And public image and private reality will now be completely separate.

There's only one thing that makes me think this won't happen. I told my vision to a hill farmer I know. 'Shucks,' he said. 'You think I could get some of those logs when the fellow's through with them? My furnace eats wood something awful.'

Tell Me, Pretty Billboard

DRUNKS TALK TO LAMP POSTS. Little girls in Victorian novels talk to imaginary playmates. I talk to the writing on packages and signs.

It, of course, has been talking to me ever since I learned to read, but it is only recently that I have started answering. Most of my life I have been a good consumer, like any other American, keeping my mouth shut and my ears open. Quietly I pushed my cart down supermarket aisles, listening to the mechanical music and being influenced by point-of-sale displays. Quietly I relaxed at home, watching teleshapes that couldn't see me, and hearing electronic voices that I could turn off but not answer. It was all perfectly normal and perfectly one-way. Never was I disturbed by my lack of I-thou relationships with vending machines, or by the fact that though a girl on a poster could stir me, there was no hope of my stirring her.

About a year ago, though, I suddenly found myself wanting to ask a cigarette machine in Burlington, Vermont, why it charged more for cigarettes than the human beings in the drug store across the street. Hardly two weeks later I had an overwhelming urge to tell an electronic voice in my living room that the 'nei' of 'neither' may have been pronounced 'nigh' by the Electors of Hanover even after they moved to England and began speaking the king's English, but the right way to say it is still 'knee.' Naturally I couldn't do either.

The urge grew, though, and before I knew it I found myself writing one letter to the company that owned the vending machine—you can sometimes find the address on the side—and another to the network that employed the rexophile announcer.

I promptly got answers from both, but the answers didn't tell me much, except that public relations is a growing field. Both might have been composed by the same man who handles correspondence for my congressman. (I had happened to send him a thoughtful letter about Far Eastern policy that same month.) All three replies had a kind of customized form letter quality that I assume is designed to give people like me the illusion that we have gotten through when we haven't, except as statistics, or what the network probably calls feedback. The congressman said he valued my opinion, and would keep it in mind. The network said it was delighted to hear from me, and its dictionary approved both the pronunciations I mentioned. The vending machine company said vending machines were very expensive, and tobacco taxes very high. It also said that as I was obviously a thoughtful person, it was sending me a questionnaire to fill out on what kinds of products I would like to see available in vending machines, and a quarter for my trouble.

It was at this point that I turned to fantasy. If dialogue with the economy has to be an illusion, I decided, it might as well be illusion of my own making. The answers would be more interesting. Since then I have been holding frequent conversations with billboards and taped messages and directions on razor-blade packages, and life is bearable again.

I'll give an example. There is a highway sign I sometimes pass in Connecticut which asserts that there are eight friendly inns on a body of water called Lake Waramaug and urges me to stay at one of them. In mere reality it would be difficult to ask that sign how friendliness can be offered as a commodity, like clean towels or room service. For the dweller in fantasy it's a snap. The last time by I had a chat with it which went something like this:

ME: Hello, you friendly sign.

SIGN: *(friendlily)* Hello-alo. That's a nice car you're driving, Mister. I like you. Why don't you take your next left and come on to Lake Waramaug?

M E : Maybe I will. Are all eight of the inns really friendly?

S I G N : Are they? Why, Mister, you don't know what friendship means till you've stayed at Lake Waramaug.

M E : *(suddenly, at 39, struck by doubts)* By gosh, you're right. I don't know. What does friendship mean?

S I G N : At our inns, it means that everybody from the bellhop to the manager is sincerely glad to see you, that we have a relaxed, friendly atmosphere–

M E : Hold it. You can't define a word in terms of itself.

S I G N : Oops. I mean a relaxed, *casual* atmosphere where everything's very informal, where the bartender smiles as he mixes you a martini, where, well, where everybody from the bellhop to the manager is sincerely glad to see you.

M E : That's great! I love big welcomes. We'll all be real friends, right?

S I G N : *(after a second's hesitation)* Sure. Every guest is a friend.

M E : Sign, this is the best news I've had in a long time. It just happens that I'm dead tired, dead broke, and very hungry. A clean bed and a good dinner are just what I need. Which of my new friends shall I stay with?

S I G N : *(icy cold)* Listen, buddy, if you're looking for charity, why don't you try the Traveler's Aid?

M E : When I've got sincere friends all around Lake Waramaug? Why should I?

S I G N : OK, I walked into that one. So they're not your real friends.

M E : But they're still sincerely glad to see me?

S I G N : Well, yes, sure. What I meant was that provided you can pay your bill, everybody from the bellhop to the manager feels sincerely glad to see you, whether they personally like you or not; and any employee that can't feel that way gets fired. Fair enough?

M E : You mean you have to be a hypocrite to work at Lake Waramaug?

SIGN: Look, I'm just a sign, and I'm ten miles back on the road as it is. Why don't you go talk to the man that wrote me?

But I don't do that. Instead I drive on up to Vermont. A few miles from home I stop to buy some groceries. (I was lying when I told the sign I was broke. Unlike hotelkeepers, I'm naturally insincere.) The first thing I buy is a package of garden-fresh frozen vegetables.

ME: What does garden-fresh mean, Package?

PACKAGE: Just what it says. What's nicer and fresher than a garden? (strikes an attitude)

> A garden is a lovesome thing, Got wot!
> Rose plot...

ME: Listen, I like gardens, too. I just want to know what garden-fresh means on a package of frozen peas.

PACKAGE: If I have to spell it out, it means the peas inside me are as fresh as if you'd gone into your own vegetable garden and picked them right off the pea vines, ten minutes before dinner.

ME: When were the peas inside you picked?

PACKAGE: Oh, about six months ago. But they've been *frozen*, Mister. They could just as well have been picked today.

ME: Were they frozen ten minutes after being picked?

PACKAGE: It was the same day. We rush them from the fields to the plant—

ME: Ten minutes?

PACKAGE: All right, if you're going to be literal about it, the peas inside me are as fresh as if you'd gone into your own vegetable garden and picked them right off the pea vines within twelve hours before dinner.

ME: And there has been no deterioration at all during the six months?

PACKAGE: I said they were frozen.

ME: Biochemical change ceases entirely when peas are frozen?

PACKAGE: It goes a lot slower.

ME: But it does occur?

PACKAGE: Look, if you don't like frozen foods, grow your own damn peas. *I* don't care.

ME: You are going to answer the question?

PACKAGE: (*in a fury*) Sure, there's a little deterioration. But it's well within Department of Agriculture Standards, and—

But at this point the woman at the check-out counter shoves the package in a freezer bag, and I drive on home. My next conversation takes place about a month later, when I'm back in New York, about to get a cup of coffee out of an automatic machine. There are no less than five choices, with a button to push for each. I can have black, sugar only, cream only, sugar and cream, or sugar and extra cream.

ME: Excuse me, but I was wondering why cream is spelled c-r-e-m-e?

MACHINE: Perfectly good spelling. The French use it, you know. *Voulez-vous un café-crème?*

ME: *Oui. Avec sucre.* But why French in New York City?

MACHINE: I didn't say it *was* French. I said the French use that spelling. As a matter of fact, it has something to do with trademarks.

ME: I don't follow.

MACHINE: Come on, you haven't grown up in America without knowing about trademarks. No company can get exclusive rights to a word unless they figure out a new spelling. That's why there are so many. You don't think serious business executive *like* to name their stuff Wonda Winda or Tas-tee Bitz, do you?

ME: Do you have exclusive rights to c-r-e-m-e? I could swear I'd seen that on some kind of make-up my wife uses.

MACHINE: Sure. That's so your wife won't think her make-up has cream from cows in it. It means a creamy or cream-type preparation.

ME: But you do mean cream from cows, don't you?

MACHINE: I mean it's fresh-made hot coffee, served 24 hours a day, which is quite a miracle, when you come to think of it.

ME: Cream from cows?

MACHINE: So it's from soybeans. You know there's a considerable body of medical opinion that thinks cream from cows is pretty dangerous stuff. Bad for the heart. You ought to be down on your knees thanking me.

ME: *(still upright)* I also know about that study of Irish twin brothers, where one brother stayed in Ireland and the other came to the United States. The ones back home used more butter and cream, but the margarine-eaters over here had more heart attacks. What about that?

MACHINE: Easy. There's lots more stresses in American life.

ME: And therefore I've got to drink soybean juice in my coffee? Well, why not say plainly that's what you serve, and why? Maybe you'd pull the trade from the dairy restaurants.

MACHINE: Mister, you think we got room for health lectures on the panel above a lousy pushbutton? C-r-e-m-e fits on the panel, everybody knows what it means, everybody's happy.

ME: But—

MACHINE: *(going right on)* What else could we do? You've got to think about customer recognition. Say we put your phrase—'soybean juice,' was it?—on the panel. Does it sound like something to go in coffee? Or does it confuse the hell out of everybody? Or maybe you want we should put 'soy sauce'? Listen, the stoops we get would be coming with plates and silverware, looking for the chow mein.

ME: But—

MACHINE: And if we put 'Cream substitute made from soybeans—better for you,' I ask you, how are we going to get it all on?

ME: But what I've been trying to say is that space is plainly not the whole problem. You've got room to help the customer recognize. Look at your own panel. I see 'White Cross Dairy.' There, right above the c-r-e-m-e. What about that?

MACHINE: *(blushing bright scarlet)* It's not very big type.

ME: Interesting phrase, though.

MACHINE: Don't say it, I know what you're thinking. To some extent you have to compromise to succeed in business, and that's a fact.

ME: By the way, what's your sugar made of?

MACHINE: Sorry, Mister. You'll have to ask the Food and Drug Administration. I just went out of order.

The Wooden Bucket Principle

April 28, 1963

T o t h e E d i t o r s : *The New Yorker*

Dear Sirs:

Under the heading 'Incidental Intelligence (Is Nothing Sa-cred? Division)' you lately reported that some Vermont farmers are using plastic bags rather than wooden buckets to catch the sap from maple trees. This is true, though plastic tubes which eliminate the need to catch the sap at all (they run it directly from inside the tree down to the sugarhouse) are more popular still. But that's not what I'm writing you about.

I am unable to decide whether you really believe that maple producers, in Vermont or out of it, were using wooden buckets until the plastic apparatus came along. Since wooden buckets started to go out not long after the Civil War ('I like tin,' said the Secretary of the Vermont Board of Agriculture in 1886), and since they had pretty well disappeared, even on hill farms such as mine, by about 1910, it seems improbable. On the other hand, city people plainly haven't been following the development of sugaring at all closely. For example, every spring your mag-azine publishes one or two cartoons showing some farmer gath-ering sap from buckets of the wrong shape, without covers, hung too high on what look to be beech trees, or possibly box elders. Most farmers do still hang buckets (they're made of galvanized

steel, with peaked lids to keep the snow out), and we mostly still boil down over wood fires, rather than oil fires, but we never tap box elders. If you can swallow those cartoons, I don't suppose you have much trouble believing in the wooden buckets.

Actually, I don't seriously blame either you or your cartoonists. I think there's something I shall henceforth call the Wooden Bucket Principle at work here. By this I mean a tendency to imagine almost anything in the country as simpler and more primitive and kind of nicer than it really is. Picture calendars are the most familiar example. Every time I see a calendar decorated with a color photograph of a New England village, I look, and I'm never disappointed. There's the little village, nestled among the hills. There's the white church. There the majestic maples. What about the filling station? It's been cropped. There are never gas stations in pictures of New England villages. Those big orange school buses don't generally get into such pictures either, nor does the town shed, with a couple of modern road scrapers lying around out front.

I also find the Wooden Bucket Principle operating in the books we have begun to buy for my daughter, who is nearing two. Some of these are books written and published in the United States in the last couple of years: animal A B C's, I-Can-Do-This-or-That books, and so forth. Supposedly they are both about and for contemporary American children. Yet I was reading from one just tonight which showed a little girl saying, 'Pick. Pick. Pick. I'm a little chick.' Behind her are five chicks, about a dozen hens, and two roosters, all wandering freely about in front of a quite charming henhouse, picking for corn. Real American chickens, of course, do no such thing, even as chicks. They neither wander nor pick. Instead they spend their time, in lots of ten thousand or a hundred thousand, locked in battery houses, never walking an inch. The cages are too small. I think their feed has Aureomycin in it.

There's a similar discrepancy in a slightly more advanced book my daughter has. This one shows a slightly older girl helping to drive in some cows. Not yet in Vermont, maybe, but on most modern dairy farms there's no place to drive cows in from.

The modern dairy cow, under the dry-lot system, lives her full life in a concrete enclosure, receiving her feed–alfalfa, mostly– from an overhead conveyor. A little girl would just get in the way on a drylot. Besides, she'd fall and skin her knees on the concrete.

I understand that in California the same method is being used for beef cattle. (One of its advantages is that permanent indoor life gives steers paler flesh. The meat thus has more room to darken, once it's cut up and put on display in a supermarket. This is handy in reducing the need to put out fresh packages.) Yet for all my daughter will learn from her books, all the cows in California hang around in the sunlight, tanned and healthy, just like the surfers.

As a matter of fact, the Wooden Bucket Principle isn't simply a country thing. I come to New York fairly often, and I've seen it in use by the Port of New York Authority. A couple of summers ago, while the second level of the George Washington Bridge was being built, the Authority started handing out explanatory leaflets at the toll booths. I have several. I well remember getting the first. It made the customary apology for the delay and gave the customary explanation that it was for my future safety and convenience. It also showed an artist's rendering of the work in hand. The bridge was in the background, looking very handsome, but somehow only three lanes wide. The foreground was taken up with sketches of two genial-looking workmen, busy widening the New Jersey approaches. They were using for this purpose a shovel and a pickax, respectively. I had only to look out the car window to see the scene in actuality: the eight lanes of traffic streaming across, the earth-movers and giant power shovels roaring about on the Jersey approaches. There wasn't a workman in sight except dimly, inside some fifty-ton piece of equipment. (I suppose bridgebuilders' flesh is paler these days, too.)

I have no idea where all this is going to end. I may still be writing letters to you when your whole magazine is put together by plastic tubes. On the other hand, I may by then have gotten into a wooden sap bucket and pulled down the lid.

Old MacBerlitz Had a Farm

AS A FORMER STUDENT of nightingales' cries, I have long known that men hear things differently. Set a few poets to transcribing nightingale notes, and one man's *jug-jug* is another man's *tereu* and a third's *whit-whit-whit*. Yes, and a fourth careful listener's *zucküt, zicküt, zidiwick, zifizigo*. If you grant even a moderate consistency to what nightingales actually sing, you are forced to conclude that the human ear operates on remarkably low fidelity.

Just how low-fi it can get, though, I didn't realize myself until I happened to stop off between planes in Singapore a couple of years ago. Most of my two days there I spent visiting a friend from college, a young Chinese scholar who lives over toward the campus of the University of Malaya. This man owns a good-sized white bulldog, and we were out walking it. It was making a spectacle of itself.

'Listen to him,' Henry said proudly. 'He'll go *wang-wang* at anything on four legs.'

'He'll bark at anything, you mean?'

'Sure. Bark. *Wang-wang-wang*. Look at that Pekinese run.'

I did look, but my mind had leaped to matters of fidelity. 'Henry,' I said, 'I want to ask you something. Have you ever heard a cow m—I mean, call to another cow?'

'Sure. Why?'

'What did it sound like?'

107

'Cows go *hou*,' said Henry. 'The outcry is especially notice-able when there's an active bulldog about. Why?'

'What about sheep?'

'Sheep go *mieh*. Why do you ask? Are you writing a book for children?

'Just curious. Goats?'

'Goats go *mieh*, too. And, for your information, cats say *miao*, and ducks *kua-kua*. Furthermore, there's a mosquito about to light on your neck, and *he*'s going *hêng-hêng*. Slap him and tell me about this book.'

I'm not writing any children's book, and I have no plans to, but it's true I've been reading a good many since I got home. I've also been sending letters to foreign farmers and consulting endless bilingual dictionaries. I am now ready to report my find-ings. They are going to come as a nasty shock to those raised on the belief that dogs the world over really do go 'bow-wow,' or that mosquitoes go 'zzz,' or that all cows 'moo,' or even that a stone drops into water with a sound resembling 'plop.' Such innocents are going to feel as lost and miserable as the Chinese nightingale in a poem Henry quoted me: 'While *ying* is its cry, seeking with its voice its companion.'

Dogs are the place to start. Dogs have no consistency at all. As far as I can tell, even Henry's bulldog hardly sounds alike to any two people in Singapore. 'That damned *gong-gong* night and day is driving me crazy,' one of Henry's Indonesian col-leagues told me, and he was not speaking of temple bells. There is a Dutch businessman in Henry's circle, and he confessed that to his Netherlandish ear bulldogs appear to say *waf-waf*. Before I left, Henry himself got interested in the vocal proficiency of his dog, and he dug up an old Spaniard who informed us that a mind shaped in Barcelona would invariably hear a bulldog bark as *guau!* Frenchmen you find everywhere, and the one I ques-tioned in Singapore said with a patronizing smile that any liter-ate man should know the verb "to bark" is *faire ouâ-ouâ*.'

Little more pattern emerged when I got home and began to consult the children's books and farmers and dictionaries. It's true that the Polish *hau-hau* does not differ much from the Latvian *vau! vau!* And I grant that all dogs in Scandinavia, irrespective of nationality, raise their muzzles on moonlit nights and let out a noise something like *vov-vov*. On the other hand, you have only to follow the path of King Ivar the Boneless from southern Norway over to Ireland to discover that the inhabitants of the former Free State have wolfhounds and terriers and even white bulldogs which bark in Irish, *amh-amh*. Swing down to Turkey, and while some Turkish dogs will set up a sedate chorus of *hav-hav*, others apparently come up yelping *kuçu-kuçu*. Read a dictionary of the sacred Pali language of Ceylon, and you find that those old Vedic Aryans called their dogs *bho-bhu-kaas*, which translates *bhu-bhu*-maker. My one real disappointment is the dogs of Japan. A bamboo importer in New York told me they bark *ming-ming*. My research shows it's not true. A Japanese dog actually barks *wan-wan*, which is much duller, and would not even faintly surprise anyone who had been to Singapore. I am only consoled by thinking of the word *pyee* in the Shangana-Tsonga language of the Bantu group; it means 'the cry a dog makes after being kicked.'

Cows are a good deal more stable than dogs. Cows have a reassuring sameness. Right there among the Shangana-Tsonga, to whose ears the sound of a gun going off–our plain English 'bang'–is best represented by the word *gibii*, even to the Shangana-Tsonga the voice of a cow seems to be saying *mhoo*. (The reply of her calf is a high-pitched *dwee*, but that's another matter.) German cows softly go *muh*. French cows elegantly if nasally bawl *meu*. Spanish cows make a noise that Spaniards spell *mu*. Russian cows say simply *moo*, and in Poland (where the winters are probably worse), it's *mooo*. Practically the only deviants I know are the versatile Hungarian cows, who call *mu-bu*; the Norse cows, who say *bø*; and the fat cows of Ireland, who, when in Gaelic mood, are able to produce a sound that their masters put into the Roman alphabet as *geim*.

Pigs are like dogs, only more so. Pigs vary. A man could travel the world over, nearly, and hear nothing even close to a friendly *oink*. Certainly not in Portugal, for example. Portuguese swine are much given to lolling in the mud, letting out little cries of *cué, cué*. In Poland, on the other hand, from identical positions in similar mud, the pigs–the younger ones, at least–call out *kwick-kwick*. Across the border in Russia, they roll their eyes and go *khru*. Not one of these words would kindle any recognition in the mind of a Hungarian swineherd, who knows very well that pigs go *röff-röff*. (Except, of course, in Japan where they go *buu-buu;* in Finland, where the cry is *snöf-snöf;* and in Italy, where the sound you hear down by the sty at swill-time is a frantic *fron-fron-fron.*)

As for sheep, the problem is to separate them from the goats. It is not easy. In China, where *mieh* is the solitary bleat uttered by either, people obviously don't even try. Nor do they in Spain, where both sheep and goats cry *bee*. And in Onitsha Province of Nigeria, it must take a keen ear to tell the *nmáa* of a sheep from the *nmée* of a goat. Things are clearer in France. There the sheep cry *bê!* and the goats *mê!* When one turns to Russia and finds that the sheep are calling *beh* and the goats *meh*, hope soars. One begins immediately to work out a theory of *b*'s to tell sheep and *m*'s to tell goats. One soon abandons it. Turkish sheep, calling out *mee-mee*, don't fit the b-for-sheep theory. Turkish goats, with their steady *bö-bö*, don't fit the m-for-goats theory. German goats, bleating *meck-meck*, fit in, but German sheep, whimpering *mäh-mäh*, ruin everything. Perhaps one had better say the hell with sheep and goats and go join the Norwegian cavalry, so as to hear the horses calling *knegg-knegg*, each to each.

I could keep on with this roster more or less forever. I want to. I want to tell how even the familiar "cock-a-doodle-doo" of a crowing rooster is not the reliable thing one supposes. (Dutch roosters, for example, stand on their dunghills shouting *kuke-leku*, which Dutch dictionaries explain as their *victoriege-kraai*, while across the border, in Germany, the roosters pronounce it *kikeriki*. I would like to report the music-hall phrase *coin-coin*, believed by Frenchmen to be the cry uttered by ducks. And the sound one hears from turkey gobblers in Poland, which is *gool-gool-gool*. It pains me that as for mosquitoes I can only mention that from pole to pole, wherever Japanese is spoken, these sober insects advance on picnics humming *bun-bun*. I had even hoped to stage a quick Iberian invasion of Norway, so as to startle those shaggy northern horses with a Portuguese cavalry squadron, every steed in it going *ri-lin-chin-chin*.

But the loud *bimbam!* and *klingklang!* of the German bells in the Lutheran church tell me that my time is growing short, an impression confirmed by the musical *cing-e-bing* coming from the little Albanian Orthodox chapel around the corner. I shall hardly have time to imagine a few coins dropping *zblunk, zblunk* into a Czechoslovakian fountain and to meditate that in Turkey those same coins would strike with a heavy sound of *cumberlop, cumberlop*. In less time than it would take a Swedish farm cart to rumble *mullra, bullra* past my window, I must be off. I'm going to Africa, on sub-safari. I want to visit the Shangana-Tsonga. If *pyhakavaka* really and truly represents the noise made in that part of Africa when a naked person falls seated into the mud, I may settle down in that onomatopoeic paradise. If it isn't, I'm coming home by way of Japan. I've always wanted to hear a nightingale that could sing *hoohokekyo*.

The Other Side

Sometimes I can hardly believe it when I read all the sad articles about how everything is breaking down in New York City. The garbage isn't collected, the air is polluted, crime is up, and the phones don't work. Pretty soon, the sad writers predict, the whole city is going to be fleeing upstate, or else to Vermont. Well, I have a counter-prediction: If the exodus ever does occur, 90% of those who flee will be back in the city within six months. At least, if they go to Vermont they will. It may be different upstate, though I doubt it.

For the first 30 years of my life I, too, was a New Yorker. Thirteen years ago I moved to Vermont. As I recall, I wanted to be in harmony with nature. I was still a bachelor then, free to quit my well-paid city job and take a respectable but much-lower-paying one up here. In short order I got married, had children, and bought an old farm to live on. I still live on it, harmonious though cold. But let's make a few comparisons between the grim life in New York City and the rural idyll in Vermont.

1. *No garbage strikes.* This is true. There has never been so much as a slowdown by the sanitation workers in my town, which I am prudently not going to name. I'll just say that this 'town' consists of 36 square miles of rolling farmland, with three or four picture-book villages dotted across it. The one I live on the edge of has 200 people.

The reason for this good behavior is very simple. There *are* no sanitation workers here (except a private one, and I'll come to him in a minute). Instead, there is a large, smelly dump to which each of the area's 2,000 inhabitants takes his own garbage. We can take it whenever we feel like it, provided we feel like it, either Wednesday or Saturday afternoons, which is when the dump is open. When I lived in New York, I worked Wednesday afternoons, and I spent Saturday afternoons prowling around the book district, or in some cool bar over a draft beer (no bars and no draft beer in this town—or in any of the ones around it, for that matter), or maybe at The Cloisters. Now I still work Wednesday afternoons, and I spend Saturday afternoons making a twelve-mile round trip to the dump with a week's garbage. And I'm supposed to feel sorry for all those poor New Yorkers who have had their collections reduced from three times a week to two?

It's true I don't *have* to take my own garbage. For $52 a year a retired farmer on the other side of town would come and take it for me. I suppose for $104 he might even come twice a week, though I don't think anyone has ever asked him to. In Vermont $104 is a great deal of money. So is $52, for that matter. Anyway, once a week is plenty. Week-old garbage hardly smells at all, except in high summer. Around here, high summer lasts about three weeks.

2. *No crowded subways*. There sure aren't any. We don't have trouble with taxis, either, since there are none. Also no buses, and the last trains were discontinued six years ago.

Suppose someone in town foolishly neglects to own a car. A child, let's say, or an old man living on Social Security, or maybe a young wife whose husband takes their one vehicle to work. Suppose this person nevertheless has the nerve to want to go somewhere. To a grocery store, say. Or to the nearest hairdresser, seven miles away. Or just to a friend's house. Well, first, he or she is at complete liberty to hitchhike on our uncrowded roads. (Once when my truck got stuck in the mud, and I had to walk three miles home, I only *passed* two cars the whole way. Neither gave me a ride. I admit it was after dark.)

Second, for only a few dollars the person who doesn't own a car can get a cab to come pick them up. The cab only has to come fifteen miles from the nearest large town in Vermont, or thirteen miles from the nearest one in New Hampshire. There is one rich old lady in town who takes a 26-mile cab trip every week to go shopping. Poor old ladies stay home. They can always watch TV. They even have two channels to choose between–three if they're one of the handful of old ladies who live on a hilltop.

We also have air service. To catch a plane, you simply go over to the regional airport in New Hampshire. I daresay the flights are canceled no more than 20% or 30% of the time. (You think we have *radar* at our airport?) Of course, even if the plane takes off, don't count on going where you expected. I'm not speaking of the hijacking that afflicts one big-city flight in every 100,000 or so. I'm speaking of failures to land because of rain, fog, snow or storms, which produce a new destination for one small-town flight in every seven or eight.

The last time I was at the airport was about a month ago. A friend had flown up from New York. As there happened to be a little ground fog, I waited three hours while he was driven back from Burlington, up near the Canadian border, where his plane got diverted without notice. Nine P M to midnight. It's true I couldn't wait at the airport, since they lock the place up around 10:30, and everybody goes home, but I didn't mind hanging around a small town in New Hampshire for three hours. It even has a movie-house–and if the last show hadn't already started, I would probably have taken in a film.

3. *We drink that pure country water, and breathe that fresh country air.* We do indeed. Of course we have to work like dogs for the privilege–but what else would we do with our time?

In New York, no one gives a second thought to water. It's just there. Why it is just there? Because a lot of nice men who work for the Board of Water Supply go somewhere upstate, throw a bunch of people off their farms, and build a big reservoir. Then a lot more nice men who work for the Department of Water Resources pipe it in to everyone's apartment or house. No tiresome meters. Every New Yorker can have all the water he wants, and it costs practically nothing.

Here, on the other hand, only a few people with artesian wells are assured of all the water they want, and for all of us it is expensive. The reason is that almost everyone in town has his own private supply. Mine, for example, comes about a quarter of a mile from a salubrious spring that I personally own (and personally dig out each spring). I also own the quarter-mile pipeline. It was a nineteenth-century lead pipe when I bought the farm, so leaky that literally no water at all reached the house. My first action after signing the mortgage was to pay $1,100 to have a new pipe laid. Guess how long that would pay for city water in New York. Five years? Twenty? Guess again. Forever is the correct answer. The average New Yorker pays 2¢ a day for water–8¢ for a family of four. He could manage that just on the interest of my $1,100, and have a good bit left over to pay those inflated taxi fares.

Mind you, for his 2¢ the New Yorker gets an absolutely reliable supply. It's so rare in New York for a building to be without water that when it does happen you can hear the screams even in Vermont. I'm not saying it's a small thing to be without water, even for a few hours. It is, in fact, remarkably unpleasant. I'm just saying that in the country it happens all the time. My spring is supposed to be one of the best in town, and I have a new pipe. All the same, three times in the last five years we have been without water. Once for twelve hours, once for a day and a half, once for four days. There is no city office to call, either, nor do plumbers like going a quarter of a mile into the woods to fix things, especially in the winter. Below zero or not, there I am down in the spring, fumbling under the icy water, while my wife shivers on the edge, passing me tools.

I haven't even mentioned droughts. There was one in the Northeast a few years ago. It caused a lot of inconvenience in New York. People weren't allowed to wash their cars, no water

was served in restaurants (unless you asked), even air condition-
ing was turned off early. We had no such petty restrictions
here. On the other hand, about a third of all the wells and springs
in town went dry. Some of them stayed dry for six months. At
the time, one of my friends, a distinguished writer in his sixties,
had his mother-in-law visiting. From Austria, by ship, and planning to stay all summer. She was about 90 and thoroughly
spoiled by the easy life in Vienna. My friend spent a good part
of the summer lugging buckets of water into the house so the
old lady could take her accustomed daily bath.

Then there's air. As I freely concede, it really *is* pure in most
of Vermont. In summer a joy to breathe. One of the reasons for
staying. Winter is another matter. At any temperature much
below zero it hurts a little to breathe. You can almost feel the ice
crystals forming in your lungs. So once again we are working
like dogs, carrying in wood to light fires to get the air warm
enough so that we can stand to breathe it. Paying for fuel oil at
higher prices than people in New York pay, out of salaries
that would define most of us as poverty-level there.

4. *In the country you get away from it all.* All what? Sweat-
ing humanity? Noise? Intrusion? I guess not. People who like
privacy had better stay in the city. You won't find it on a farm.
You may be out of sight of the next house, and sometimes there
may be only two cars down the road all morning, but private
you're not.

In the first place, you're not out of hearing. And I don't mean
of those nice lonely 'somewhere-a-dog-barked' country sounds to
be found in old-fashioned novels. I mean of modern technology.
For one thing, your neighbors all have chainsaws with which
they're busy cutting wood in order to keep from freezing to
death. The sound of a chainsaw carries about a mile, besides
echoing from hillsides. Sometimes you can hear two or three at
once. If you imagine several dentists working in your mouth
simultaneously, you will get some notion of the effect.

Then there are several seasons of the year when invasion of
your land is perfectly lawful. Summer is not one of them, and

New Yorkers who have a 'place' they go to from mid-June to Labor Day do not encounter this phenomenon. But as soon as they have gone back to the sanctuary of their apartments, invasion time rolls around for us. The first wave, of course, is called hunting season. So many hunters from Massachusetts and Connecticut—well, they're not all exactly *hunters* but, rather, so many men with guns—come barreling up that just recently a bill was introduced into the Vermont Legislature to limit the number here at one time to 40,000. It is not expected to pass. When the hunters get here and have parked their campers on your land, they sally out to shoot. Deer are the official target, but a fair number of cows, horses, goats and human beings also drop each season. Lest you think I exaggerate, I should mention that a child's pet goat was shot in this very town last fall, and a farmer in the next town was shot and killed the fall before.

On the first morning of deer season you look out over your hundred acres, and between the parked trucks and the moving figures with loaded guns and fluorescent red jackets, you can hardly see the landscape. You can hear the rifle shots, though. Later you can pick up the broken glass and the cans.

You are free, of course, to post your land against hunters. People try this every year. Many years ago I tried it myself. But the law is written in such a way that legal posting is a lot of work—there must be a sign every hundred feet, visible at such-and-such a distance—and the penalties for trespassing on posted land are slight. Furthermore, posting annoys the hunters, and they usually tear the signs down right away. What most Vermonters do is simply keep their pets, children, and selves out of the woods during hunting season. Even if they own the woods.

Hunting season lasts only a few weeks. But the next invasion, which is called snowmobile season, lasts all winter—that is, from the first of December to the end of March. To some extent the same people who were up here shooting bullets at you during deer season are now back roaring their snowmobiles across your fields. Every Friday you see hundreds of cars heading north on

Interstate 91, each with a little trailer behind it and on the trailer two snowmobiles. But most of the snowmobilers are your fellow Vermonters. When I was a boy in New York, the standard middle-class criticism of the poor was that in front of the tenement you saw a row of shiny cars, and on its roof a giant T V aerial. The feeling was that people who live in tenements should spend their money on things they don't enjoy so much. The standard criticism of poor Vermonters now is that they shouldn't have so many snowmobiles.

They do, though, and they love them. Night riding is especially popular. One trail from the village goes about 50 feet behind my house, through the orchard, then down over a stone wall and onto my neighbor's land. One night last March, sixteen snowmobiles, lights blazing and engines throbbing, came down that trail between 10 and 11 P M . I couldn't have run out and stopped them, even if I wanted to, because I can't possibly run, snowshoe, or ski through deep snow as fast as they can snowmobile. I could put up signs, which I think would be respected by about four-fifths of the snowmobilers, but that four-fifths includes many of my neighbors and several of my friends. No point in spoiling their fun when the other fifth, including the teen-ager who used his snowmobile to cut everyone's wire fences winter before last, would be coming anyway. (Why did he cut the fences? Pure esthetics. A pasture full of clean snow, unsoiled by snowmobile tracks, offends him the same way a freshly painted wall offends a small child with a crayon.)

Snowmobiles, in case you've never heard one, make the same sort of noise as chainsaws. Not an honest roar like the subway— to be on a local platform when two expresses pass through at once is like listening to Beethoven by comparison—but a rising and falling whine. If mosquitoes were six feet long and powered by gasoline engines, they would sound like snowmobiles. And in case you have never had one in your yard, they are sudden death for any plant foolish enough to stick up through the snow, such as the top of a rose bush, or a newly planted apple tree.

Spring, at the moment, is secure from invasion. But wait! Before you decide you can stand being shot at for a few weeks and snowmobiled around for a few months if you can only have a beautiful country spring, alone with the wildflowers, let me remind you of the trail buggy, the all-year, all-terrain vehicle. It hasn't fully caught on yet, but it's about to. Then there won't be any more wildflowers. Except in places like Central Park and the Bronx Botanical Gardens.

5. *People are so honest in the country.* There seems to be a general feeling that country people never steal, whereas in the city the time that your neighbors have left from ripping off your apartment they devote to stealing your car. The second part of this statement may be true. The first is city myth. It is true that there are few spectacular robberies in this little town—but that's only because there is nothing spectacular to steal. What we do have, people take, including stuff a city thief just wouldn't bother with. A couple of years ago, for example, some men came with a large truck and stole about 50 feet of stone wall from in front of a house just up the road. This past winter someone stole nearly all the chains that summer people like to put across their driveways before they leave in September. Weathervanes off barns are another popular item.(To be fair, it is widely believed that employees of Boston antique dealers take these.) But the regular business, the bread-and-butter thievery around here, is breaking into summer people's houses, which is always done in the winter when the residents are away. Since a lot of their houses are on unplowed back roads, for the thieves it used to mean either hard work on snowshoes or the risk of getting your truck stuck there until spring. Now the thieves dash in by snowmobile, whisk around the driveway chain (if still present) and are away in a trice. I admit no present snowmobile can accommodate major items of furniture.

6. *The country is full of charming wildlife, while the city is infested with rats.* It is certainly true that there are more chipmunks in Vermont than in New York City. But there are also more rats, mice, bats, blackflies, toads, and snakes.

Rats, for some reason, figure as the ultimate horror of city life. Talk of rats is good for a shudder on Sutton Place any day. Live in the country, though, and you can do more than shudder. You can, for example, take a .22 along to the dump and stay for a happy hour of rat shooting. This sport is not allowed in my town, but in the next town over, the Saturday afternoon rat jamboree is one of the big social events of the week. Not that we feel deprived. We get our rat fun at home. My wife once shot a rat right in our barn. That, however, was not nearly as exciting as the time she opened the feed bin to get grain for our two beef cattle, and five startled rats dove out on all sides of her.

I could easily go on with some discussion of blackfly season or an account of rural telephone service, but the way I want to end this piece is not with more drawbacks of the country, but an advantage of the city.

This past winter I was in New York for a week, catching up on movies, museums, ballet, draft beer, and so on. While I was there, I kept hearing this tempting ad on the radio for a Czecho-slovakian restaurant. To me, already that's exotic—you can easily guess how many East European restaurants there are in Vermont. When the ad went on to say that this particular place had been chosen by the senior food critic of the *New York Times* as the very best of all the Czech restaurants in New York, I could have broken down and cried. We hardly get a choice of dough-nut stands in Vermont; New Yorkers idly pick and choose among the Czech restaurants.

Now it is obvious that Continental restaurants serving game dishes are only for the well-to-do, which not all New Yorkers are—though a hell of a lot more New Yorkers than Vermonters. But on that same visit I was down on Irving Place, and wandered into the kind of lunch counter every New Yorker takes for granted: three middle-aged men behind the counter, and 30 kinds of delicious pastry on it. I didn't get any of that. I just put down my 65¢ and got a bowl of the kind of lentil soup every New Yorker also takes for granted—full of little carrots and some kind of superior wurst, and with it, completely casually, a great length of better bread than I've had in thirteen years in the country. A man has to like chipmunks an awful lot before he moves away from that.

POSTSCRIPT. This article was written in 1972. Several things have changed since then. For one, snowmobiles are less of a menace than they used to be. Local snowmobile clubs all over the state have begun to police maverick riders; the deliberate breaking of fences is now a rare thing. The trail that used to go behind my house has been moved back into the woods, and we are no longer wakened by night riders. Better yet, trail buggies have still not materialized. If the energy shortage continues, perhaps they never will.

We had another drought in the summer of 1977, and this time nearly half the wells and springs in town went dry. A lot of people drilled artesian wells in a hurry. One of my neighbors now has one 800 feet deep—at a base charge of $7 a foot, plus $4 a foot for the casing, plus $275 for the pump, and so on. Three or four blockfuls of New Yorkers could pay their water bills on the interest of what he spent.

A new airline now flies in and out of our foggy little airport; and while canceled and rerouted flights are still common, they are much less common than they were in 1972. We also have train service again. Amtrak restored it four years ago. The trains are practically never canceled or rerouted—I can think of only once, after a major flood.

They have a good many other drawbacks, though. What Amtrak gave us was one southbound train a day, and one northbound. Neither is what you'd call convenient. Going south, you board the train at 1 AM, and then jolt down toward Massachusetts at 25 miles on hour, eventually reaching New York City in the morning. Going north to Montreal, you are out waiting on the icy platform at 3:50 AM (Or if you are coming home from a trip to New York, it is at this hour that you are ejected into the Vermont night.)

But there is a rumor that someday the train will travel as fast as it did 30 years ago (no one even dreams that it will get back to its old nineteenth-century speeds); all it would take would be track maintenance to the tune of what three or four miles of interstate highway would cost. If that happens, the rumors say, we may be granted a day train.

The Saturday afternoon rat jamboree at the dump has vanished. So has the dump. We now have a sanitary landfill. Rats on the farm are holding their own nicely, though. And someday we may have a bat jamboree. There are lots of bats in Vermont. Increasing numbers of them are rabid. A doctor we sometimes see at church suppers is fond of saying flatly that a tenth of all the bats in Vermont are rabid. He adds cheerily that so far only three human beings are known to have survived the bite of a rabid bat. If offered the choice of a bat-bite in Vermont or a mugging in New York, I might pick the mugging.

The bill to limit out-of-state hunters to 40,000 never passed. But the quantity of broken glass and beer cans in the woods has declined anyway. Most people give the credit to the new Vermont law which requires a 5¢ deposit on bottles and cans. If New York, New Jersey, Connecticut, Rhode Island, Massachusetts, and New Hampshire would also pass such laws (Maine has one already), our woods might be downright clean.

But the biggest change since 1972 is that highway culture has arrived. The town I live in is unhit so far. Two towns away, however, there is a stretch of land where the Pizza Hut jostles the Burger King and both are near the Holiday Inn, from which a moderately good shot with a .22 could knock out windows in the Howard Johnson's and the Sheraton Motor Inn.

This town is vulnerable. An interstate highway goes through it, and there is an exit. At the moment, it must be one of the ten prettiest exits from an interstate highway in New England. All four corners are open farmland and woods. Two farmhouses and one red barn are visible.

Any of the four corners would, of course, make a dandy site for a Pizza Hut or a filling station or a two-hundred-room motel. People who like the town rural are acutely aware of this; and two years ago, when it looked as if the owner of one of the corners might be forced to sell, there was some talk of a group in town trying to buy the land cooperatively, so as to keep it the way it is. The trouble is that that would simply make the other three corners even more valuable as development properties. One corner would now be closed to competitors—indeed, it would constitute a scenic amenity, which the developer would regard as put there for his personal convenience. Even if some friendly millionaire bought all four corners and endowed them as farmland in perpetuity, a developer could merely move back a few hundred yards and set up on the edge of this now really splendid park. Short of abolishing interstates (or developers) no one has thought of a solution.

FIRST PERSON RURAL

has been set in Linotype Janson, an old style face first issued by Anton Janson in Leipsic between 1660 and 1687, and typical of the Low Country designs broadly disseminated throughout Europe and the British Isles during the seventeenth century. The contemporary versions of this eminently readable and widely employed typeface are based upon type cast from the original matrices, now in the possession of the Stempel Type Foundry in Frankfurt, Germany. The book has been printed and bound by Haddon Craftsmen.